From My Journaling Beta-Testers

"We really enjoyed these!"

"I remember doing pages and pages of dull equations with no creativity or puzzle-thinking, but now as a homeschool mom, I'm actually enjoying math for the first time! My daughter's math skills have skyrocketed and she always asks to start homeschool with math."

"Thank you for a great intro to Playful Math!"

"All of the kids were excited about their journals. My oldest kept going without prompting and did several more pages on his own."

"We had a lot of fun doing your math prompts. We had never done any math journaling before, but we will certainly integrate this into our weekly routine from now on."

A PLAYFUL MATH SINGLE

312 Things To Do with a Math Journal

GAMES, NUMBER PLAY, WRITING ACTIVITIES, PROBLEM SOLVING, AND CREATIVE MATH FOR ALL AGES

Denise Gaskins

Tabletop Academy Press

© 2022 Denise Gaskins
Print version 1.0
All rights reserved.
Except for brief quotations in critical articles or reviews, the purchaser or reader may not modify, copy, distribute, transmit, display perform, reproduce, publish, license, create derivative works from, transfer or sell any information contained in this book without the express written permission of Denise Gaskins or Tabletop Academy Press.

Tabletop Academy Press, Boody, IL, USA
tabletopacademy.net

ISBN: 978-1-892083-61-6
Library of Congress Control Number: 2021924569

DISCLAIMER: This book is provided for general educational purposes. While the author has used her best efforts in preparing this book, Tabletop Academy Press makes no representation with respect to the accuracy or completeness of the contents, or about the suitability of the information contained herein for any purpose. All content is provided "as is" without warranty of any kind.

Contents

Preface: Cats and Math 1

Writing to Learn Math 5

1: What Is a Math Journal? 7
2: Making It Work 16
3: Thinking, Writing, and Thinking About Writing 24

The Journaling Prompts 37

4: Games 38
5: Number Play 51
6: Geometry 64
7: Math Art 75
8: Writing 86
9: Freewrites 96
10: Explanations 102
11: Research Reports 110
12: Measurement and Data 118
13: Problem-Solving 128
14: Experiments 137
15: Create Your Own Math 149

Conclusion 165
   16: Continue the Adventure 166

Appendixes 173
   21 Favorite Online Resources 174
   Quote and Reference Links 177
   Acknowledgments and Credits 186
   Index 188
   About the Author 197

This is the wonderful thing about just thinking and playing with half-formed thoughts: often exciting ideas will flash into your brain when you least expect them.

—James Tanton

Preface: Cats and Math

In all the books I write, my goal is to encourage families to explore the world of math in a new way. To enjoy thinking and playing with ideas. To delight in the beauty of numbers, shapes, and patterns. And to fight for true understanding, doing whatever it takes to help math make sense for our children.

With that "fight for understanding" on my mind, back in January 2021 while the Covid-19 pandemic raged on, I launched my first math journaling Kickstarter project, called "Make 100 Math Rebels."

To my surprise, my daughter's cats Cimorene and Puck signed on to lead the Kickstarter promotions. Because cats know the Internet, and they know how to make people do whatever they want.

Or at least, that's what they told me.

Cimorene thought everyone should order one of the big sets of three paperback or hardcover journals. Books come in boxes, after all, and boxes are important to cats.

Puck agreed that boxes are a good thing. But he thought people should get into journaling in any format they liked. Puck values curiosity and creative thinking, and math journaling is all about teaching students to explore ideas and think creatively about math.

The Kickstarter succeeded beyond expectations. More than a hundred parents and teachers signed on to help me create three beautiful journals for adventurous students, with full-color, parchment-style pages that make writing fun. Along with the journals, I offered supporters a checklist of one hundred math journaling prompts to help draw out their children's mathematical thinking.

Months later, I've tripled that original list into a full book with more than three hundred ideas to spark creative, liberal-arts mathematics. As I'm wrapping up work on this book, the cats are still plotting ways to spread the news about writing to learn math.

Cimorene worries that many children (and their parents) struggle with a fear of math. She thinks that's because school math can seem stiff and rigid. To children, it can feel like "Do what I say, whether it makes sense or not."

That's a horrid feeling. It reminds Cimorene of being trapped in the carrier bag for a trip to the vet. She wants everyone to know that math journaling is not like that. In fact, journaling makes number play fun like catnip.

Nobody wants a trip to the vet. Cimorene hopes you'll take her advice and try a math journaling prompt instead.

But Puck thinks most people are confused by the idea of math journals.

Cats know how important it can be for students to explore math and try new things. Playing with ideas is how kittens (and humans) learn. Many people understand that children need to do hands-on experiments in science. But Puck believes that most adults don't know how to do a math experiment.

The Cat Escape Puzzle

To show how your children can experience the joy of creative reasoning, Puck decided to create a puzzle about saving cats from their mortal enemy.

Imagine the dog ran into the kitchen, so the cats need to get off the floor. There are three chairs around the table. There are two cats, and they don't like to share a seat. How many different ways can the cats jump onto the chairs?

When Puck was a little barn kitten, his mama taught him that the best way to learn is to figure things out for yourself. So he won't give you the answer to his puzzle.

Your children may draw pictures, write explanations, or use equations. They can work alone or with a friend. When someone finds an answer that makes sense to them, and their friend can't find anything they missed, that's good enough.

And then the fun begins. The real point of a math experiment is to change something in the problem and see how that changes the answer.

What new Cat Escape Puzzle will your children create? What if there are four chairs, or three cats, or only one cat? What if there are more chairs? What if there's only one chair? (A math horror story, from Puck's point of view!) How might the puzzle change if the cats were willing to share a seat?

What questions will you ask?

The Princess Puzzle

Cimorene refuses to let Puck have all the math journaling fun. She wants children to understand that there are many approaches to solving any math problem, so she created a new cat puzzle of her own.

> *The Princess of Cats has a luxuriously soft tail about 12 inches (30 centimeters) long. Her tail is three times the length of her noble head. Her beautiful, furry body is as long as head and tail together. How long is the Princess from her delicate nose to the tip of her majestic tail?*

What can children do with a problem like this? They may want to make a list of the things they know from the story. Perhaps they will draw a picture of the cat and label the proportions. Each will take their own approach to figure it out.

And then the best part of any math journal prompt is when kids make their own math. Will they write a puzzle about their own pet? Or about their favorite animal? Encourage your children to be math makers, sharing their creations with their friends and family.

As every cat knows, learning is a lifelong adventure that everyone can enjoy.

Listen to Cimorene and Puck, and help your children explore their own ideas about numbers, shapes, and patterns through journaling. I hope your family has as much fun playing with these prompts as the cats and I had writing them.

—Denise Gaskins, with Cimorene and Puck
Rural Illinois, September 2021

Section I

Writing to Learn Math

Mathematics is not about following rules. It's about playing and exploring and fighting and looking for clues and sometimes breaking things.

Einstein called play the highest form of research. And a math teacher who lets their students play with math gives them the gift of ownership.

Playing with math can feel like running through the woods when you were a kid. And even if you were on a path, it felt like it all belonged to you.

Parents, if you want to know how to nurture the mathematical instincts of your children, play is the answer.

What books are to reading, play is to mathematics.

—Dan Finkel

Mathematics as a liberal art: In this 16th-century engraving by Gregor Reisch, Lady Arithmetica generously shares her wisdom.

When I was in front of the class demonstrating and explaining, I was learning a great deal, but many of my students were not. My definition of a good teacher has since changed from "one who explains things so well that students understand" to "one who gets students to explain things so well that they can be understood."

— STEVE REINHART

Chapter 1: What Is a Math Journal?

ONCE UPON A TIME, MATHEMATICS was considered a liberal art—an important part of any well-rounded education. Artists painted images of the angelic ladies Arithmetica and Geometria sharing their wisdom with human scholars.

Somehow, over the centuries, math lost its connection both to wisdom and to art.

Now, too often, the school math curriculum forces students on a relentless treadmill from kindergarten to calculus. Our test-driven culture rewards a fast memory and leads children to believe that "math" means cramming facts and procedures into their heads so they can perform on demand.

It's no wonder many kids grow up thinking they're no good at math.

And far too many parents feel unable to help their children learn. They worry about their kids falling behind, which raises the stress level to the point of tears. Mom and Dad are frustrated. The child is discouraged. Doing math homework feels like stumbling through an emotional minefield.

How can we help our children step off this treadmill and rediscover the liberal art of mathematics?

The problem is, we're all a product of our own schooling. Just as we are hoping to shape our children and their future through training them, we were shaped by our own childhoods. And for most of us, our schooling gave us a totally wrong idea of what math is all about.

School and society teach us to view mathematics as a race. You run as fast as you can from one topic to the next. You must get the answer quickly. You need to follow instructions and score high on tests, and then you win. Or if you don't, you're a loser.

But let me give you a new vision of mathematics. I want you to think of math as a nature walk. There's a whole wide, wild world of interesting things—more ideas, more patterns, more concepts than you and your children would ever have time to study. And everywhere you look, there's something cool to discover.

In his book *Measurement*, math teacher Paul Lockhart compares doing math to a jungle safari:

> "*Mathematical reality is an infinite jungle full of enchanting mysteries, but the jungle does not give up its secrets easily. Be prepared to struggle, both intellectually and creatively.*
>
> "*The important thing is not to be afraid. So you try some crazy idea, and it doesn't work. That puts you in some pretty good company! Archimedes, Gauss, you and I—we're all groping our way through mathematical reality, trying to understand what is going on, making guesses, trying out ideas, mostly failing.*
>
> "*And then every once in a while, you succeed … And that feeling of unlocking an eternal mystery is what keeps you going back to the jungle to get scratched up all over again.*"

If you explore this mathematical world with your children, you're not behind. Wherever you are, you're not behind, because *there is no behind*. There's only "We're going this direction," or "Let's move that way," or "Hey, look what I found over here." If your children are thinking and wondering and making sense of the math they find, they're going to learn. They're going to grow.

The key to helping our children have success with math is to focus on teaching the real thing. Real math is about making sense of ideas.

Real math is about creative reasoning.

School math rewards children who follow directions, even though it's tedious to memorize stuff that you really don't understand. And to always follow someone else's rules, that's boring. But to figure out things for yourself can be exciting.

When you embrace this adventure of learning math through playful exploration, you'll be surprised how much fun thinking hard can be. It doesn't matter whether your students are homeschooled or in a classroom, distance learning or in person. Everyone can enjoy the experience of playing around with math.

In this book, I'll teach you one of the best ways I know to put real math into practice and help children experience math as a nature walk: math journaling.

Recording Their Mathematical Journey

In a math journal, children explore their own concepts about numbers, shapes, and patterns through drawing or writing in response to a question. Journaling teaches them to see with mathematical eyes—not just to remember what we adults tell them, but to create their own math.

All they need is a piece of paper, a pencil, and a good prompt to launch their mathematical journey. The prompts in this book include number play, math art, story problems, mini-essays, geometry investigations, brain-teasers, number patterns, research projects, and much more.

My journaling prompts invite students to take any rabbit trail that interests them and discover whatever they will, without worrying about grades, testing, or state standards. Everyone can enjoy journaling because creativity is fun. And when children get a chance to be creative in an area they normally think of as drudgery, it feels like a refreshing treat.

Through journaling, children come to realize that learning is more than memorizing facts and procedures, and they develop a richer mathematical mindset. As they explore their own thoughts, they begin to see connections and make sense of math topics. They grow confident in their ability to think through new problems.

When students write about what they're learning, they build deeper layers of understanding. The process of wrestling ideas into words forces them to pin down nebulous thoughts and decide what they really believe. Journaling gets children actively involved in their own learning. They are more likely to remember what they learn when they write it down.

For children who struggle with numbers and abstraction, writing offers a more familiar way to grapple with concepts. It helps them see themselves as mathematical thinkers.

For students who find math easy, writing reminds them that there's more to being good at math than just getting the right answers. And for those who struggle with words and language, writing about math can feel more natural than many language-based writing assignments.

Math journaling can help you as a parent or teacher, too. If you want to know what your students understand about math, their writing gives you a glimpse into how they are thinking. Some teachers use journal writing as an "exit slip," asking students to jot down a sentence or two about each lesson before leaving class.

Five Types of Journaling Prompts

In Section II of this book, I've organized the math journaling prompts into twelve categories, which may seem overwhelming at first glance. Here's a simpler way to classify the prompts by the type of reasoning involved.

1: Game Prompts

Game prompts break through the idea that math is dull and boring. They help students develop a positive attitude toward math while practicing their number skills or strategic thinking.

For example:

Basic Nim (two players): Draw 10–15 circles (called "stones"). On your turn, mark out one or two of the stones, removing them from play. Whoever marks the last stone wins the game.

Game prompts can also serve as fodder for the other types of prompt questions. We might ask students to analyze the mathematics of the game, to determine whether either player has an advantage, or to explain how they make strategic decisions during game play.

2: Content Prompts

Content prompt questions deal with the concepts of math and the topics studied. They can range from a short summary of a recent lesson to an in-depth research report on math history. Or they may pose a number-play puzzle or a word problem for students to investigate.

For example:

Choose any base number and investigate its powers. For example: If you choose a base of three, the powers are $3^1 = 3$, $3^2 = 3 \times 3 = 9$, $3^3 = 3 \times 3 \times 3 = 27$, $3^4 = 3 \times 3 \times 3 \times 3 = 81$, etc. Extend the list as far as you can. What patterns do you see in the powers of your base number? What other questions can you ask?

Content prompts help students see the bigger picture of a topic. Too often we teach by breaking a math topic into small, bite-size chunks. But writing helps students to step back and put all those little pieces in perspective.

3: Artistic Prompts

Artistic prompts encourage children to express their creativity in playing with mathematical designs. The prompt may propose a geometric or numerical constraint for the artwork. Or it may be open-ended, allowing the students to choose their own responses.

For example:

Use dotty graph paper. Connect dots to create an eight-sided shape. Are all the sides of your octagon the same length? How can you tell? What kind of design can you make with octagons?

Artistic prompts inspire children to make mental connections in a way that abstract number problems can never do. Students feel the relationships of angles and lines as they draw a shape. And these prompts may lead to informal geometry proofs, like determining whether the sides really are the same length.

4: Process Prompts

Process prompts explore and explain the way a student solves problems. They ask learners to organize their ideas and reflect on their problem-solving strategies. Process prompts involve *metacognition*, which means "thinking about your own thinking."

For example:

Describe a mistake you made in math, or a problem you missed on a quiz or test. What went wrong? How will you avoid this error the next time? Do you understand the problem now, or is there something more you need to learn about it?

Process prompts help students recognize their own understanding. Too often, math class is about learning to follow other people's thoughts, not about thinking for ourselves. But students already have many ideas about math, and the best way of teaching is to draw out and strengthen those ideas.

5: Affective Prompts

Affective prompts ask about the student's feelings and attitude toward mathematics. This includes self-assessment: How is your math comprehension growing? What is easy for you, and what is most difficult?

For example:

Have you heard that your brain keeps growing the more you use it? And that mistakes help you learn even more than when you get things right? How do these scientific discoveries affect your attitude toward math?

Affective prompts support students in relating math to their own personal experience. They make math seem more "real" to students, more relevant to things they care about, more meaningful. Writing helps students take ownership of their math experience.

And One More Type: Quotation Prompts

When you're looking for ways to prompt student writing, short quotations can be a great resource. I love quotations: Everything I might possibly want to say, someone else has already said it better than I ever could.

You can share one of your own favorite quotes or search for a new quip online. You may want to sample the tidbits on my blog's "Math and Education Quotations" resource page.[†]

Short-Response Prompts

Let students choose how they want to react to the quotation. Or offer one of the following questions:

- What did the author mean? Put the thought in your own words.
- Do you agree or disagree? Why?
- Is it a general principle, or only for specific situations? Describe a time when the quote might apply, or when it might not.

---

† *denisegaskins.com/quotations*

- Tell a time in your life when you lived up to the quotation—or when you wish you had.
- How does the quote relate to math, science, history, or another subject?

Research Prompts

Short exercises are great writing practice. But occasionally you may want to assign deeper essay topics, such as:

- Look up the author's name online. Who are/were they, and why do people care what they said?
- What have others said about the same topic? Search out a variety of quotes related to this one. How are they similar? How are they different?
- Does thinking about the quotation make you want to change anything, in yourself or in the world? How could you put that idea into action?

This Is Not a Lesson

Math journaling is different from the normal process of learning math. A typical school math book asks questions where the teacher always knows the answer. This turns math into a performance subject in which our children are constantly being judged. Some students enjoy the chance to show off their knowledge, while others feel like failures.

But journaling prompts ask questions for which we adults do not know the answer because the topic gets filtered through each child's own mind. Students come to a task at their own level and explore their own ideas. Everyone may learn something different, but they all grow as mathematicians.

The journal prompt is not a lesson to be learned. Even with the research prompts that require a student to seek out new information, there is no specific thing we want them to see. It's more like a directed nature exploration: "What can we find hiding under this log?" We're building awareness, helping them see that there's more to mathematics than they realized.

The journal prompt is not a quiz to be graded or an essay to be judged. Even when a prompt has one specific right answer (which is rare), its primary purpose is to draw out each student's own creative reasoning. How they approach the problem is much more important than whether they figure out an answer.

Our role as parents and teachers is to listen to the children. We want to hear their ideas and understand what's going on inside their minds. When we ask about their own thoughts, our children are the experts. And that's an enormously powerful feeling.

I used to think my job was to teach students to see what I see. I no longer believe this. My job is to teach students to see; and to recognize that no matter what the problem is, we don't all see things the same way. But when we examine our different ways of seeing, and look for the relationships involved, everyone sees more clearly; everyone understands more deeply.

—RUTH PARKER

Chapter 2: Making It Work

DO YOU WANT YOUR CHILDREN to see math as a liberal art that focuses on creative thinking? Imagine the freedom of no longer worrying about speed and memory. Instead, they can dive headlong into the deeply refreshing waters of reasoning and problem-solving.

To take the plunge into math journaling, you'll need three things: a plan, a few supplies, and a playful math mindset.

Let's take them one at a time…

What's Your Plan?

Think of this book as a cookbook for mathematical discovery. Each prompt is like a recipe that your family or class can try, and these recipes are gathered into chapters with various categories of foods for thought.

As with any cookbook, you don't need to start at the beginning and work your way through to the end. Instead, skip around. Try an assortment of prompts from different sections to give your children a balanced math diet.

Just as not everyone likes the same foods, so not every student will enjoy the same prompts. Like recipes, math prompts are merely guidelines, so feel free to adapt the question to your child's taste. If an

activity flops, set it aside and try something else another time.

Be patient. Children need time to get comfortable with writing about their thinking. Putting thoughts on paper is harder than it sounds. For students who are new to writing about math, you may want to limit journaling to once a week at most. Many homeschooling families plan regular lessons four days a week and use Friday for enrichment activities. This would be a great way to work in journal projects, perhaps accompanied by a game or a math storybook from the library.

If that sort of playful enrichment "uncurriculum" sounds good to you, check out the variety of math games for different ages in my *Math You Can Play* series of books or on my blog. For book recommendations, look at the Living Math website.[†]

Journal entries may be short or long, depending on the student's level of interest. Many prompts take only five to fifteen minutes, but others can keep a child going for hours. A research project or math experiment may stretch over several days. Don't rush through the activities, but follow your children's lead.

[†] *tabletopacademy.net/math-you-can-play*
denisegaskins.com/math-games
livingmath.net

Also, your students' comfort with writing will vary tremendously. For young children or reluctant writers, a few words or sentences may be all they can handle. On the other hand, some students might fill up a whole page and still have more to say.

Most of the prompts in this book are designed for repeated play. Some of them have blank spaces, where you can choose a number or math concept for that day and a different one the next time around. Others are open-ended investigations or opinion questions that will seem fresh again as the student grows in understanding over time.

Gather Your Supplies

There is no "right" way to do math journaling. Students may use any bound notebook or loose paper, lined or unlined, or graph paper of any type you have on hand. For written prompts, some students may prefer typing on the computer.

Personally, I love dotty graph paper for journaling because I can start a line anywhere on the page, and the dots serve as anchors for drawing shapes or patterns. My favorite paper has dots arrayed at 1/4 inch or 0.5 centimeter. Young children may want wider spacing: 1/2 inch or 1 centimeter. Triangle dot paper (isometric grid) is also fun, because it encourages writing at different angles.

The Incompetech website is a great place to download graph paper of all varieties.[†]

If your students are using a bound journal, you may want them to draw the geometry and math art prompts on blank paper. They can use the journaling page to record what math they see in their design and how they thought about creating it.

In particular, geometric constructions made with a compass and straight-edge (or a ruler) are much easier to draw on a loose sheet of plain paper. For best results, use masking tape to hold the paper in place so it doesn't shift under the compass.

[†] *incompetech.com/graphpaper*

In addition to your journaling paper, you will find the following supplies useful on your mathematical adventure:

- pencils, both plain and colored
- colorful gel pens
- a ruler for making straight lines
- a drafting compass for drawing circles and comparing distances
- other drafting tools, like plastic triangles or a circle template
- dice for playing games
- a deck of playing cards, poker or bridge style

Less common, but useful for a few of the prompts:

- colorful construction paper
- scissors
- glue or clear tape

Fix Your Own Mindset

There is an old saying that if the only tool you have is a hammer, every problem looks like a nail. When the only math tool we have is the rote procedure we learned in school, then that's what we hit our kids over the head with. Whatever their struggle in math, our response can only be, "Follow these rules because that's how to get the answer."

But if we want our children to develop the ability to reason creatively and figure out things on their own, then we need to master some new teaching tools. Or not actually new, for humans have used these tools throughout history.

Help your kids practice slowing down and taking the time to fully comprehend a math topic or problem-solving situation with these classic tools of learning: *See. Wonder. Create.*

Learning Tool #1: See

Pay attention. Focus your mind to see what is there (not just what you expect). Take turns saying things that you see. Write a list. Spring off each other's comments to find even more specific bits or pieces you had overlooked.

Look carefully at the details of the numbers, shapes, or patterns you see. What are their attributes? How do they relate to each other? Also notice the details of your own mathematical thinking. How do you respond to a tough problem? Which responses are most helpful? Where did you get confused, or what makes you feel discouraged?

Learning Tool #2: Wonder

Look at the problem or concept through the lens of your own curiosity. Begin a new list and start asking questions. Let the brainstorming side of your mind run wild.

Ask the journalist's questions: who, what, where, when, why, and how? Who might need to know about this topic? Where might we see it in the real world? When would things happen this way? What other way might they happen? Why? What if we changed the situation? How might we change it? What would happen then? How might we figure it out?

Learning Tool #3: Create

Apply your imagination to the math and begin to make it your own. Investigate the questions you saw and wondered about. Pursue new ideas.

Create a description, summary, or explanation of what you learned. Make your own related math puzzle, problem, art, poetry, story, game, etc. Or create something totally unrelated, whatever idea may have sparked in your mind.

Math journaling may seem to focus on this third tool, creation. But even with artistic design prompts, we need the first two tools because they lay a solid groundwork to support the child's imagination.

When parents and teachers use these learning tools, we must quiet our own thoughts to really hear what our students are saying. And in the process of listening, we help them learn how to reason through a problem: how to harness the things they know, whatever they do remember, to figure out a concept or situation that they don't yet understand.

Mathematical thinking is a habit of mind that you and your children can learn and grow in.

To see how these tools work, let's go back to the metaphor of math as a nature walk. Imagine that you are hiking through the woods and your children find a new plant. You don't immediately pull out a field guide to look it up. Or if you know the plant, you don't just say its name and keep walking—not if you really want the kids to remember what they learn. Instead, you and your children focus on the plant itself. What is the shape of the leaves? What's the color? Do you see a stem, flowers, seed pods? Where is it living? Does it seem to like shade or full sun, dry soil or soggy? What else can we see or guess about the plant? You may even take time to make a sketch of the plant with colored pencils or drybrush watercolor.

After students discover as much as they can on their own, then they're prepared to understand new information from a book.

You can do the same with a math problem or a journaling prompt that has your children stumped. Don't immediately rush to a math book and look for answers. Or if you know the answer, don't just give it away—that only spoils the fun. Instead, focus on making sense of it. Ask what your students notice about the math. After they pay attention to the things they can see, start to ask about what they wonder, and pose some wonderings of your own. As you and your kids follow through on some of these wonderings, you'll all gain insight into what's going on in the problem.

Eventually, the children will see a path forward to creating their own foundation of understanding. And then they'll be prepared to make sense of what they find in a math book.

Where Are the Answers?

In Section II of this book, you'll find twelve categories of journaling prompts to get your students writing about math. Not every prompt will suit every student, but children of all ages can find something to enjoy.

There is no answer key to these prompts. The point is to explore the world of math, not to find one particular solution.

Mathematicians use the things they know to figure out things they don't know. Math is never about finding the One Right Answer to a problem. Real math is a way of thinking, of reasoning about things you know and things you want to figure out.

After all, it took more than three centuries for mathematicians to find "the answer" to Fermat's Last Theorem—but in struggling with the question, people developed all sorts of new mathematics they might never have thought about otherwise.

Some of these journal prompts will leave students in a similar position. They may wonder about the math without reaching any conclusion. Or they may find a partial solution, but there will still be more about the topic to discover in the future.

Other prompts have a definite answer for children to find, but there is always more than one path to that answer. Math journaling helps students discover the path that makes sense to them.

In either situation, it's up to the students to decide when they are finished with the question. If they do reach a conclusion, teach your kids how to check their results: "Does my solution make sense? Is there a way to confirm it? Could I approach the problem a different way and still get the same answer?"

This process of looking back and testing our work is an important part of mathematical problem-solving. If students develop this habit now, it will serve them well in higher math classes and in adult life.

And sometimes, like Andrew Wiles (who finally solved Fermat's puzzle), students may later realize their original answer wasn't complete or there were factors they hadn't considered. Feel free to revisit the problem at that time and see what else you can discover together.

When we view math as a nature walk, there's always something new to explore.

As you range across the mathematical terrain, some places will be easy walking, where children can run ahead if they like. Other paths may be steep and rocky, or low and boggy, reducing your progress to a painful slog. In the next chapter, we'll look at two rough spots that may cause trouble on your math journaling adventure: the struggle of writing, and the problem of how to respond to your student's work.

If this feels hard, that doesn't mean you're a failure. It means you're doing the right thing to get better.

—BEN ORLIN

Chapter 3: Thinking, Writing, and Thinking About Writing

MOST PEOPLE FIND IT DIFFICULT to put their thoughts into words. So, it should not surprise us when our students complain that math journaling is hard.

At any age, children who are unfamiliar with the challenge of pinning down vague ideas often claim they have nothing to say. Even the strongest students may balk at writing about math: "I just *do* it. I can't *explain* it."

Julie Bogart, author of *The Writer's Jungle*, describes five stages of growth in a child's writing skill:

- ♦ JOT IT DOWN (early elementary grades): The child talks, and you write the words down.

- ♦ PARTNERSHIP WRITING (upper elementary grades): You and the child work together to express the child's thoughts.

- ♦ FALTERING OWNERSHIP (middle school): The child fluctuates between needing support and working independently.

- ♦ TRANSITION TO OWNERSHIP (early teen years): You gradually step aside as the youth grows in skill and confidence.

♦ ACADEMIC WRITING (later teen years and beyond): The young adult continues to grow in understanding and begins to participate in (or at least to eavesdrop on) the Great Conversation of thinkers and writers throughout history.

Don't try to rush your child through these stages. For a struggling writer, set small goals: one phrase, one sentence, twenty words, or one short paragraph. Proficient writers, too, may have days when the mind feels blank, when small goals offer an encouragement to get started. Just as the physical body must grow at its own pace, so must your child's math and writing skills.

For more information on each learning stage, visit the Brave Writer website.[†]

Tips for Brainstorming Math

When your students struggle to express themselves, encourage them to talk through their thoughts before trying to set them on paper. Work together, because learning to communicate is a social skill. And most children find talking easier and more natural than writing.

Take notes on what they say. Then let them copy those notes in their journal, if they wish, or use them to launch their own writing.

Coax your children to expand on their ideas: "Tell me something else you know. Let's make a list of words you might use." Try making a concept web, with lines connecting related topics and branching off into new possibilities. Or start by drawing a sketch. Then explain the picture with words.

Sometimes it helps to begin with a simple subject-verb sentence. Then add details. Use adjectives and adverbs. Add a subordinate phrase. Make the sentence come alive. Write more sentences to expand the idea into a paragraph.

When students get stuck in the middle of their writing, ask them to read aloud what they have written so far. Think of yourself as a

---

[†] *bravewriter.com/getting-started/stages-of-growth-in-writing*

facilitator, not as a teacher. You want to draw out their thoughts, not fill their mind with yours. Ask questions. Tell them which parts make sense to you and where you feel confused.

Teach them to cycle through their writing, going back over it to add details and clarify explanations. Show them how to use transition words that express the flow of logic: first, then, so, because, therefore, etc.

Metacognition Takes Practice

Thinking is natural, but meta-thinking is a learned skill. For most humans (child or adult), it takes time and practice to get comfortable with writing down our thoughts.

One way to get started is to have children summarize what someone else has written. This lets them practice putting thoughts into words without the creative burden of coming up with the ideas.

As students grow more aware of their thinking process, they will be better able to communicate their ideas. Teach them the word "metacognition." Having a name for this process of thinking about their own thinking helps students remember how to do it.

If students have trouble recognizing the complexity of their own thoughts, talk about the process of making a familiar decision. How do they decide what to wear, or what game to play with friends, or which book to bring home from the library? Brainstorm a list of the factors that might affect their decisions, such as what the weather will be, which games they have played recently, whether the book is for a school assignment or for free reading.

Often, humans remain completely unaware of how our mind works through the day. We pay no attention to the details of daily life. If we had to stop and think about each tiny movement of putting on our shoes, for example, we'd never go anywhere.

But when we pay attention to our own thoughts (including the false starts, dead ends, and loops of circular reasoning), we begin to understand the act of thinking itself. This helps us to think more clearly and not to give up when things are hard.

Remind students that there are no wrong answers in math journaling. We are delving into the world of math and probing our own minds. Whatever we discover is valid. There may be times when a later, closer examination makes us change our minds, and that's okay, too. Learning is an ongoing process.

Responding to Student Journals

Math journal prompts offer a wide range of options for students to explore. Most of the prompts do not have a "right" or "wrong" answer. Our goal is to root around in some small corner of the world of math, to lift a stone and peek underneath it, just to see what we can find.

The idea that being good at math means finding the right answers is a huge myth. Of course, many problems in math do have a single right answer. But even for those problems, the answer is not the real math of the problem. Math is all about thinking.

It's like taking a road trip. You may have a destination, but there are many paths you could take to get there. Different students may take different paths—they may think about the problem in different ways. It's this reasoning that is the real math, and the right answer is just a side effect of reasoning well.

Our goal in the journaling activities is to get students thinking about the different paths they can take in any problem, so they can explore what interests them.

Let's look at an example. Consider this geometry prompt:

Draw a large circle. Draw its diameter. Connect both ends of the diameter to any other point on the circle. Do it again, connecting the diameter to another point, and another. What do you notice? What questions can you ask?

Young children will find the construction itself a challenge. They may be surprised to see a triangle appear. Depending on its shape and orientation, they may not consider it a proper triangle. After all, in most picture books, triangles are usually equilateral and almost always rest flat on a side.

Young students might also see that the diameter of the circle is longer than either of the triangle's other sides. But it's not as long as both of them together.

Elementary and older children will discover they can create many quite different triangles following these instructions. They may wonder which of these triangles is the largest. They might guess (correctly) that it's the one with the third corner point as far as possible from the diameter, but they may have no idea how to prove that fact.

They may also wonder what fraction or percent of the circle is contained in the triangle, though answering that question could be beyond their skill.

Middle school and older students might wonder about the angles produced by this construction. They may notice the smallest angle is opposite the shortest side, and the largest angle is opposite the circle's diameter. Some may even manage to prove that the largest angle measures 90° (a right angle) for all possible triangles.

Or they might go in a completely different direction, deciding to add more lines to their construction and create an artistic design.

The point of the journaling exercise is not to discover any specific piece of mathematics. Rather, we want our children to play with the ideas and explore whatever connections they find interesting.

Hierarchy of Editing Comments

Because journal writing is so personal, students are easily overwhelmed by criticism. If you make too many comments, your student may feel like you are trying to seize control of their expression.

Build your students' confidence by pointing out whatever you admire in their work: clear statements, vivid word pictures, clever phrasing, imagination, and so on. Then limit yourself to at most one or two critical comments. Focus on meaning, on understanding what the student is trying to communicate. Help them make that as clear as possible.

Honor the student's work by not marking up their paper. If you must make corrections, put them on a sticky note. Leave the child in control of which changes to incorporate into their writing—or whether they want to let the paper stand as it is, a record of their understanding at the time.

Consider these areas for improvement, beginning with the most important.

1. Purpose

Did the student make the central idea, experience, or theme clear? Is this primarily an exploratory piece? Or is it written to inform a reader? Does the student need to summarize, analyze, or argue their point?

2. Development

Does the piece hang together? Can the reader easily follow the student's logic? Did the student use evidence, examples, data, etc. to support the explanation or argument?

3. Audience (optional)

To whom is the student writing? For most journaling activities, the audience is the student's own self, along with perhaps a teacher or peers. If the piece will be shared with a wider audience, the writer may need to put extra effort into defining terms and elaborating on ideas.

4. Readability (even more optional)

If the piece will be shared, does the student need to proofread for language mechanics?

It's easy to find mistakes in spelling, grammar, or punctuation. But these areas have little to do with mathematical understanding. If we want our students to think about the math, then that's where we need to focus our attention as well.

If you find it impossible to ignore the minor distraction of spelling or punctuation errors, ask your students to read their writing to you. That will allow you to focus on the higher levels of conceptual understanding.

As teacher and author Marilyn Burns explains, "Writing in math class isn't meant to produce a product suitable for publication, but rather to provide a way for students to reflect on their own learning and to explore, extend, and cement their ideas."

Journal writing is a first draft, not a finished product. Even professional writers often ignore the mechanical details in a first draft. They focus on organizing and expressing their ideas. The grammar and punctuation can be fixed later, before they submit the piece for publication. Proofreading and clean-up is always the final step, after all the real thinking is done.

Or Try a Dialogue Journal

Your student may prefer an alternate approach to math journaling. In a dialogue journal, the child and adult alternate writing messages for the other to read. The child may ask a question, give an opinion, or react to a recent lesson. Then you might offer an answer, praise an insight, or encourage further investigation.

This process of give-and-take discussion may extend a math prompt over several days, allowing for greater depth of reasoning as you investigate the topic together. Math education is not a race. Take time to enjoy the journey of discovery.

A dialogue journal provides yet another tool in mentoring your children on their lifelong adventure of learning. And it can be loads of fun.

Miscellaneous Tips and Tidbits

The journaling prompts coming up in Section II are organized into twelve categories: Games, Number Play, Geometry, Math Art, Writing, Freewrites, Explanations, Research Reports, Measurement and Data, Problem-Solving, Experiments, and Create Your Own Math.

Within each category, prompts are collected in "Task Card Book" sets of four or eight according to the crazy numbering system explained below.

I've tried to arrange the puzzles within these sets in order of increasing difficulty. If you have a young child, try the first prompt of each set and save the later prompts until they grow into them. On the other hand, older students may find those first prompts too easy and prefer to skip ahead.

My Weird Numbering System

In the beginning, I collected math prompts to go along with the journals in my Math Rebel Kickstarter. I made a list of one hundred prompts, and I promised that if we hit a stretch goal, I'd format them as two booklets of printable task cards.

It turned out that four cards fit perfectly on a page, so I bumped the list up to fifty-two prompts per booklet (thirteen task-card pages) and sent them off to everyone who bought a journal.

After the Kickstarter project ran its course, I kept tinkering with those prompts: cutting some, tweaking others, revising and adding new ideas. And organizing them into booklets of fifty-two: four prompts in each category, plus four extra "Create Your Own Math" prompts. I ended up with six booklets, which are now available in the Printable Activity Guides section of my publisher's online store.[†]

When I decided to compile this math journaling book for my *Playful Math Singles* series, I knew it made sense to put all the similar prompts together, no matter which task card book they came from.

[†] payhip.com/tabletopacademypress

I considered numbering them straight through, from 1 to 312 in order. But in the end, I decided to keep the numbering system as it was.

So, if you've purchased one of my *52 Math Journal Prompts* task card sets, you'll find the prompts in here numbered to match your cards. If you want to look up a specific prompt, you can search by number or by card title.

And if you don't have any of the task card books—well, you can just roll your eyes at my weird numbering system. I'll understand.

Join the Math Rebellion

If you missed my Math Rebel Kickstarter, you may be curious: What exactly does a "math rebel" do?

Math rebels play and have fun with math. They write any true answer except what the textbook expects. Math rebels make the answer as crazy as they like.

For example, if the textbook answer is 57, a math rebel might write:

$$100 - 43$$
$$\text{or } 2 \times 5^2 + 7$$
$$\text{or } {}^{120}\!/\!_2 + (-3)$$
or

"The total number of mushrooms in the basket, if three hobbits each picked nineteen 'shrooms (not counting the ones they ate)."

As your students explore the journaling prompts or work through their normal homework, they may enjoy answering questions math-rebel-style.

When math rebels get a math worksheet or homework page, they don't start writing straight away. First, they examine the page to see if the problems look familiar. Do they know what the teacher or textbook wants them to do? Math rebels always care about the truth. So first, they learn what the problem means and how to figure it out. After they know how to solve the problem, then they can start working on their creative answer.

Teach your students to live by the two rules of the math rebellion:

(1) You are allowed to write anything that is true.

(2) You are not allowed to write anything that is not true.

Those are the most important rules of mathematics. Anything else is just advice, and math rebels only follow advice when it makes sense to them.

A Bit of Vocabulary

Grids and Arrays

If the math prompt says to *make a grid*, that means to draw squares in rows and columns. A grid is like graph paper, but we usually need the squares to be big enough to write in. If you have dotty graph paper, the dots provide guidance for drawing consistent grid squares.

If the math prompt says to *make an array*, that means to draw dots in rows and columns. If you are using dotty graph paper, you may simply outline a border around the number of dots you need. Often, the prompt will call for connecting these dots according to certain rules. You might want several copies of the dot array on your paper, so you can experiment with variations.

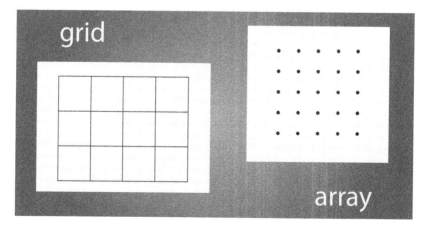

Geometric Shapes

A *polygon* is a flat shape made with straight lines (*sides* or *edges*) that meet at corner points (*vertexes*). Polygons are always *closed* shapes, which means they connect all the way around.

Quadrilaterals are four-sided polygons. You are probably familiar with squares and rectangles, but you may also want to consider parallelograms, trapezoids, kites, arrowheads, and irregular (random-looking) quadrilaterals.

A "kite" is a *convex* (all angles bump out) quadrilateral, and a "dart" or "arrowhead" is *concave* (an angle bumps in). Both have two adjacent short sides, two adjacent long sides, and one line of symmetry.

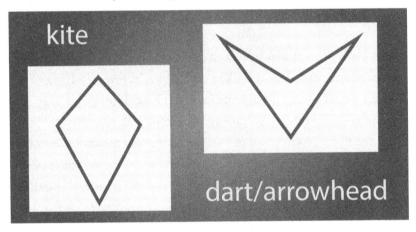

Pentagons are five-sided polygons. *Hexagons* are six-sided polygons. *Heptagons* have seven sides. *Octagons* have eight, *nonagons* nine, and *decagons* have ten sides. *Dodecagons* are fun polygons to play with, having twelve sides in all. Keep in mind that any of these may be irregular. Hexagons, for example, include all six-sided polygons, not just the ones that look like honeycomb cells.

Circles have a *circumference* (the distance around) and a *diameter* (the longest distance across, through the center). *Pi* is the ratio of circumference to diameter, which means how many times would you have to walk across the circle to go the same distance as one time around it.

$$Pi = circumference \div diameter$$

Circles also have a *radius* (the distance from the center to the edge, or half the diameter). And *chords*, which are any straight lines across the circle from one point on the circumference to another. A *tangent* is a line that barely touches the circle, intersecting at only one point. (Polygons can also be tangent to a circle, and circles to each other.)

The Most Important Tip

Have fun with the math prompts. If an activity doesn't make sense or feel playful to your students, move on to something else. You can always try again another day.

I encourage you to participate alongside your children. Math often works best as a social endeavor, whether we bounce ideas off each other or simply work at the table side by side.

Play. Discuss. Notice. Wonder. Enjoy.

Section II

The Journaling Prompts

The question, "What did you learn?" implies the process has ended. This coming school year, I will characterize learning—for myself and for students—in the following way:

Learning is having new questions to ask.

If I have learned something, it is because I can ask questions that I previously could not.

See, in math classes, asking questions is usually a sign that you have not learned. "Any questions?" is a signal to students to speak up if they don't get what has just been explained.

We have it all backwards.

It shouldn't be, "What questions do you have?"

[I hope you have none so that I can tell myself you learned something.]

It should be, "What new questions can you ask?"'

[I hope you have some because otherwise our work is having no effect on your mind.]

—Christopher Danielson

Games help set the culture I want to develop: Teaching students that multiple approaches and strategies are valued; trying is safe; and conversations about why, how, and discovery are the goals.

—JOHN GOLDEN

Chapter 4: Games

YOU CAN PLAY SOME GAMES right on your journaling page. For others, use the page to keep score and to make notes about game strategy.

Games are the ultimate re-playable activity prompts. As children repeat a game, they try variations on their previous moves to gain extra advantage. This sort of experiment mirrors the approach a mathematician may take when faced with a problem. What if we try this, or that? How do things change, and what stays the same?

After your child masters the ordinary version of a game, try a *misère* variation. In a misère game, the move that otherwise would win now makes you the loser. Students must reconsider their strategy and think more deeply about the game.

Think of other ways to modify the game rules. What if students changed the number of cards to draw, or how many dice to throw? If the game uses dice, can they figure out a way to play it with cards or dominoes? Or transfer it to a gameboard? Or is there a way to use money in the game? Or can they change it into a whole-body action game? Perhaps using sidewalk chalk?

Older students may want to analyze a game. Does one player have the advantage, or do both players have an equal chance of winning? What's the best move? Can they find a strategy to increase their odds? How are fairness and randomness linked?

Task Card Book #1

—1—
Bowling

(solitaire)

Draw circles in a bowling-pin pattern. Write the numbers 1–10 in the circles. Roll two dice and cross out any combination of circles that exactly matches that sum. Those are the pins you knocked down. Roll again, trying to hit more pins.

If all the numbers left are 6 or less, you may choose to roll only one die. Can you knock down all the pins?

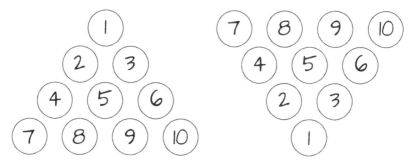

Bowling pins are arranged in a triangle and numbered from the shortest row to the longest. The triangle may point up or down, whichever you prefer.

—2—
Basic Nim

(two players)

Draw 10–15 circles (called "stones"). On your turn, mark out one or two of the stones, removing them from play. Whoever marks the last stone wins the game.

Misère variation: Whoever marks the last stone loses.

—3—
Sam Vandervelde's Criss-Cross

(two players)

Draw 3–8 dots spaced out like the corners of a large polygon. Add up to 7 more dots inside the polygon.

On your turn, draw a line (or curve or squiggle) between two currently unconnected dots. If two dots are already connected, you may not draw another line between them. Lines may not cross. The player who can't move loses.

For more information on the game, including teacher's notes and discussion questions, see the Math Teachers' Circle website.†

—4—
Secret Number Codes

(two players)

Each player chooses four secret numbers less than 20.

$$A = \_\_, B = \_\_, C = \_\_, D = \_\_.$$

Take turns with a friend asking for algebra clues like "What is $A + B + C$?" or "What is $D \times A$?" The first to guess the other players' code wins. Or just play until you solve both codes.

---

† *mathteacherscircle.org/session/game-criss-cross*

Task Card Book #2

—53—
2-D Nim

(two players)

Draw a rectangular grid of 10–20 squares in area. For example, a 3 × 5 grid makes a 15-square gameboard. On your turn, cross out one or two of the grid squares—but you can mark two squares only if they share a side. Whoever marks the last square wins.

Misère variation: Whoever marks the last square loses.

—54—
Pig

(solitaire or group)

Roll a die as many times as you want, adding the numbers to your score. Stop when you wish, and pass the die to the next player. Beware: If you roll a 1 before you stop, you lose all the points you added during that turn. The first player to reach 100 points wins.

—55—
Blockout

(two players)

Play on square graph paper (lined or dotty). Each player needs a different color pencil or pen. On your turn, roll two dice and multiply them. Color one completely connected shape with that area.

If you can't make a shape that fits on the page, you miss that turn. The game ends when there are two missed turns in a row. Whoever colored the greatest total area wins.

—56—
Sonya Post's Substitution Game

(solitaire or group)

Write a simple equation at the top of your paper. On your turn, write on a new line. Copy the equation from the line above, except replace any one number with an equivalent expression. For example:

$$2 + 5 = 7$$
$$2 + 8 - 3 = 7$$
$$2 + 8 - 3 = 14 \div 2$$
$$2 + 8 - 3 = (100 - 86) \div 2, \text{etc.}$$

Task Card Book #3

—105—
One Hundred Up

(two players)

The first player names any number from 1 to 10. On each succeeding turn, the player adds any number (1–10) and says the new sum. The player who reaches 100 wins the game.

—106—
Connect 4

(two players)

Draw a 10 × 10 grid of squares or use graph paper. Each player chooses a symbol: X, O, your initial, a small star, etc. Take turns "dropping" your symbol into a column by marking the lowest open square. Whoever gets four of their own symbols in a row (any direction) without gaps wins the game.

—107—
Chomp

(two players)

Draw a "candy bar" grid at least 3 × 3 squares in size. Mark the top left square as poison. Players take turns eating the candy.

To take a bite, you choose any grid square and mark it out. You must also mark all the squares to the right and below that square, as if you broke off a whole chunk of the candy bar. Each bite must consist of whole grid squares—no fractional bites allowed!

This method of eating squares will push the players up and to the left. Whoever is forced to take the final square loses the game.

The dark player claims more imaginary candy but loses the game. All that matters is the final poison square.

—108—
Sim

(two players)

Draw a circle. Add 6 dots spaced out around the circumference. Each player needs a different color pencil or pen. Take turns drawing straight lines across the circle to connect two of the dots. The first person to complete a triangle in their own color *with all three corners on the circle* loses the game.

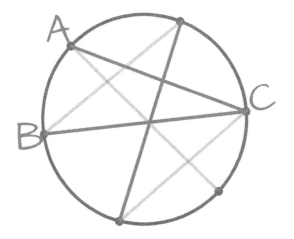

Dark gray's small triangle near point C doesn't matter. But if dark gray connects A to B, they'll lose the game.

Task Card Book #4

—157—
Double Digit

(solitaire or group)

Take turns rolling a die. Add that number to your score as either "ones" or "tens." For example, if you roll 3, you can choose to add 3 or 30. After everyone has seven turns, the player whose score is closest to 100 wins.

Optional challenge: Anyone who goes past 100 loses.

—158—
Traditional Nim

(two players)

Draw three or more large rectangles, called "piles." Within each pile, draw several circles, called "stones." Give each pile a different number of stones.

On your turn, mark out one or more stones from one pile, removing them from play. You may mark any number, up to the whole pile.

Whoever marks the last stone of all loses the game.

—159—
Tsyanshidzi

(two players)

Draw three or more large rectangles, called "piles." Within each pile, draw several circles, called "stones." Give each pile a different number of stones.

As in Nim, you can mark out one or more stones from one pile, up to the whole pile. Or you may choose to mark out the *same* number of stones from *all* the piles.

Whoever marks the last stone of all loses the game.

—160—
Gomoku

(two players)

Gomoku is like a boundless game of tic-tac-toe on graph paper. Each player chooses a symbol: X, O, your initial, a small star, etc. Take turns marking your symbol in any unclaimed grid square. But three-in-a-row is too easy with such a large gameboard. To win, you must get five of your marks in a row with no gaps.

Task Card Book #5

—209—
Place Value Nim

(two players)

The first player writes down any decimal number. On your turn, subtract at least 1 from the digit in any place value column, but not more than the digit itself. Name the amount you are subtracting: 3 tens, or 7 thousandths, or whatever. Write the new number. The player who reaches zero wins.

—210—
Greedy Pig

(solitaire or group)

Roll two dice as many times as you want, adding the numbers to your score. Stop when you wish, and pass the dice to the next player. Beware: If you roll a 1 before you stop, you lose all the points you added during that turn. If you roll double-1, your score resets to zero. The first player to reach 100 points wins.

Optional variation: If you roll doubles other than double-1, you have to roll again. You can't end your turn on doubles.

—211—
Make a Square

(two players)

Draw a 6 × 6 or larger grid of squares, or play on graph paper. Each player chooses a symbol: X, O, your initial, a small star, etc. Take turns marking your symbol in a grid space. The first player who marks four symbols that form the corners of a square wins the game.

Winning squares do not have to line up with the grid. But if your square is at an angle, you must convince the other players it's a true square.

Misère variation: Whoever forms the first square loses.

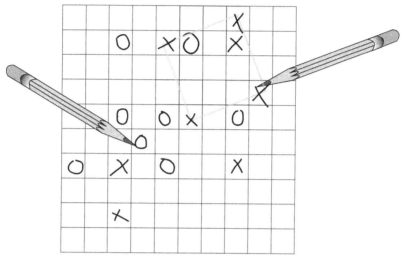

The O player's last move was a good strategic play. Can you see why? Unfortunately, the X player claims a winning square. Do you agree?

—212—
Wythoff's Nim

(two players)

Draw a large grid of squares for your gameboard, or play on graph paper. Mark the lower left-hand corner of the grid with a star.

The first player marks an X (or crown shape) in any square of the top row or farthest-right column. This mark represents a chess queen whose goal is to capture the star.

Each player in turn moves the queen any number of squares in a straight line to the left, down the board, or diagonally, and makes a mark at her new position. The queen moves always in the direction of

A chess queen can move any number of spaces in a straight line, but she cannot jump over another piece. When the queen moves into a square with an opposing piece, she captures that piece.

the star, never away from it.

The player who moves into the final square, capturing the star, wins the game.

Task Card Book #6

—261—
Jenna Laib's Skip-Counting Game

(two players or small group)

Agree on a number to skip-count by, a starting point, and a target number. For example, "Count by threes from 11 to 47."

Each player in turn chooses to write one, two, or three skips of the counting-by number. For example, the first player may write "11, 14." The second player might add "17," and then the next move could be "20, 23, 26." And so on.

The player who reaches or passes the target number wins.

Extra challenge: Skip-count by a fraction, decimal, or negative number. Or even by algebraic expression. High school students, can you count by $3x - ⅓$?

—262—
Ben Orlin's Row Call

(two players)

Draw a tic-tac-toe grid. Each player chooses a symbol: X, O, your initial, a small star, etc. On your turn, your opponent names a row or column that has at least one open square. You choose where in that row or column to make your mark. The player who gets three in a row in any direction wins the game.

Variations: Play on a 5 × 5 grid and try to get four in a row. Or use a 6 × 6 grid, trying for five in a row.

—263—
Avoid Three

(two players)

Draw a 3 × 3 or larger grid of squares, or play on graph paper. Take turns marking X in any square. All players mark X. Watch out for marks in a row, even with gaps in between. If your mark makes three Xs that can be connected by a straight line, then you lose the game.

—264—
Walter Joris's Sequencium

(two players)

Draw a 6 × 6 grid of squares. Each player needs a different color pencil or pen. Players begin by each writing the number 1 in opposite corner squares. Then take turns writing numbers until the grid is full.

On your turn, choose a square that touches any of your own previously numbered squares at a side or corner. Write the number that is one greater than the square you are touching. Draw a short line to connect those two numbers.

If your new square is touching more than one of your numbers, continue counting from the higher number. Whoever reaches the greatest number wins the game.

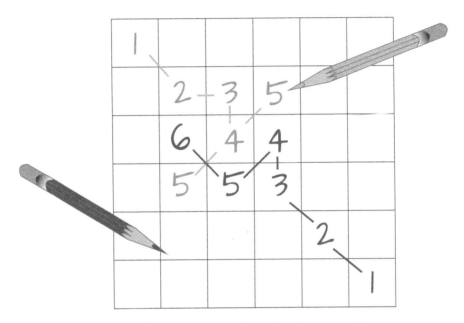

Your number sequence may branch and may cross the opponent's line. Each branch continues the counting sequence from whichever square it touches.

Math is not just adding, subtracting, multiplying, and dividing. It is the mystery of numbers within numbers and the discovery of how numbers keep changing the world.

—Savannah Sanders

Chapter 5: Number Play

Number play doesn't have to follow school math methods. Remember the math rebel rules: A student may write *anything* that is true or that makes sense.

Most number play prompts offer nearly infinite variation. Change the numbers in the description, and wherever there is a blank you may put in any number you like. Each time you revisit the puzzle, it's new again.

Older students may experiment with fractions, decimals, or exponential notation. Or try numbers in another base—do the patterns they found hold up when they change the way they count? Can they express these patterns with algebra?

Task Card Book #1

—5—
Five Cards Make 10

Turn up five ordinary playing cards. (Ace = 1, Jack = 11, Queen = 12, King = 13, Joker = 0, and number cards count at face value.) Use one or more of these numbers to write a math expression that equals 10.

You can use any math operations you know. Then find another way to make 10.

How many different ways can you say "10" with the numbers on your cards?

—6—
Multiplication Wheel

Draw a circle and write any numbers you wish around the edges, like the numbers on a clock. Draw spokes out from the center, dividing your wheel into one *sector* (slice-of-pizza shape) for each number.

Extend the spokes out to reach a second, larger circle. In this circle, write the double of each sector number. On the next time around,

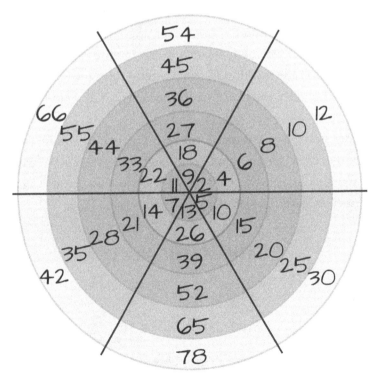

The multiplication wheel is like a round times table, but it can start with any numbers.

write the triple of each sector number, etc.
How large will your multiplication wheel grow?

—7—
Number Sums

Write the counting numbers on a single line, with a bit of space between them. On the next line, write another row, so that each number is the sum of the two numbers above it. Keep writing new rows of number sums. What patterns do you see in the numbers and their sums? What other questions can you ask?

—8—
Triangular Numbers

You've heard of square numbers. Triangular numbers are their smaller cousins. Arrange dots in a bowling-pin pattern. Count the dots to find the triangular number: one dot in the first row ($T_1 = 1$), two in the second row ($T_2 = 1 + 2$), three in the third row ($T_3 = 1 + 2 + 3$), etc.

Keep adding more dots. Each row is one dot longer than the previous row. How many triangular numbers can you find? Do you see any patterns? Can you think of any questions to ask?

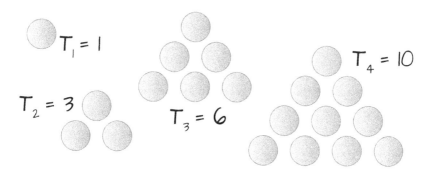

The first four triangular numbers are 1, 3, 6, and 10.

Task Card Book #2

—57—
Double Trouble

Write any number. Then write its double. Then double that new number. Keep on doubling. How high can you go before you run out of room on the page?

—58—
Numberstorming

Numberstorming is brainstorming about numbers. Write down everything you can about the number ____. You may include arithmetic expressions, expressions with words ("3 less than 10"), number properties (odd, prime, and so on), and other information ("days in a week").

Take turns with a friend adding more ideas to your numberstorm.

—59—
Make a Million

Draw six boxes in a row, each big enough to write a digit from 0 to 9. Below that, draw six more boxes:

☐ ☐ ☐ , ☐ ☐ ☐
☐ ☐ ☐ , ☐ ☐ ☐

Draw cards or roll dice to get random numbers. Write each number into one of the boxes. (For 10s and face cards, write a zero.) Once a number is written, you can't move it. When all the boxes are full, add your two six-digit numbers.

How close did you get to making one million? Compete with a friend to see who can get closest.

—60—
The Power of a Pattern

Choose any base number and investigate its powers. For example: If you choose a base of three, the powers are:

$$3^1 = 3$$
$$3^2 = 3 \times 3 = 9$$
$$3^3 = 3 \times 3 \times 3 = 27$$
$$3^4 = 3 \times 3 \times 3 \times 3 = 81, \text{etc.}$$

Extend the list as far as you can. What patterns do you see in the powers of your base number? What other questions can you ask?

Task Card Book #3

—109—
Open Number Line

An open number line is a small section of the whole, infinite line of numbers. It can start and end wherever you like, which makes it a handy tool for thinking about numbers.

Draw an open number line. The midpoint of the line is ___. One endpoint is ___. What number is at the other end? What other numbers can you find on that section of the number line?

[Choose numbers that fit the student's level: simple numbers, big numbers, fractions, mixed numbers, decimals, negatives, algebraic unknowns. Or let students pose their own number line puzzles.]

—110—
Broken Calculator

You need to calculate _____, but the "__" button on your calculator is broken. Can you use your calculator to find the answer without doing the math by hand?

Examples: Figure out 43 × 41 with a broken "4." What buttons would you push? Or calculate 7.5 ÷ 6 with a broken "5."

Extra challenge: Have more than one broken key.

—111—
Dan Finkel's Super-Broken Calculator

You have a calculator with only two buttons, "+7" and "–3." [Or choose any two math operations.] Whenever you turn it upside down, the calculator resets to zero. How many different numbers can you make it display? Are there any numbers you can't make? How do you know?

—112—
The Year Game

Use only the four digits in this year's number to make math expressions. Or use the four digits in your birthday year, or the month and date of your birthday. You must have all four digits in each expression, with each digit appearing only once. You may use any math operation you know: +, -, ×, ÷, brackets, etc.

For example, using the year 2021: 2 × 0 × 2 × 1 = 0, and 22 + 10 = 32. Can you calculate all the numbers 1–10? 11–20?

What other numbers can you make?

Are any numbers impossible to make with your four digits?

Hint: Perhaps these answers are possible, but not with the math your children have learned so far. In mathematics, a "this is impossible under these conditions" proof is a real achievement. So if a student can show for sure there's no way to make 13 with the digits 2-0-2-1 using third-grade math, that's a mathematical win.

Task Card Book #4

—161—
Scrambled Math Facts

Draw a square or rectangular grid at least 4 × 4 in size, with room to write in each grid space. In the top left-hand corner, write + or × for which operation you want to practice. Write any numbers in any order along the top row and left side column. Challenge yourself to practice with any numbers, not just the standard math facts for 1–10.

Can you fill in the math fact answers for all the grid squares?

For an additional challenge, try your hand at the scrambled times table puzzles on Iva Sallay's FindTheFactors blog. Those puzzles give a few products in the grid, and you need to figure out which factors go where in the top row and left-side column.[†]

FIND THE FACTORS 1-10   Level 1

×									
				4					
			1						
						100			
				49					
	64								
36									
		25							
				16					
		81							

FIND THE FACTORS 1-10   Level 2

×									
								27	
									18
								63	
									6
								9	
									27
								45	
					50	20	10	60	
						56	64	32	

—162—
Subtracting Reverses

Choose any two-digit number AB. Reverse the digits to make BA. Subtract the smaller number from the larger. If the answer is another two-digit number, repeat the process until you get down to a single digit.

† *findthefactors.com*

Try another two-digit number, and another. Do you notice a pattern? Is it always true, or do some numbers break the pattern? What other questions can you ask?

—163—
Number Snakes

Make a "snake" out of circles (large enough to write in) connected by lines. On each line, write a math operation like "+7" or "÷2." Then write a number in the head of your snake.

In each circle after the head, write the answer to the calculation on the line between. Or trade snakes with a friend and see whether you can stump each other.

Extra challenge: Make an algebra snake by putting an expression like "$3x + 1$" into the head. See Don Steward's blog for examples.[†]

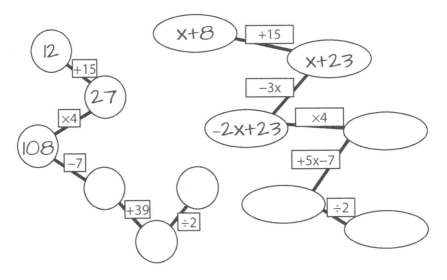

You can use the same operations to make an arithmetic or algebra snake, or you can use variables in the algebra operations. Either way, the algebra snake will need bigger spaces to write in.

[†] *donsteward.blogspot.com/2018/01/algebra-snakes-and-branches.html*

—164—
Is It a Pattern?

You know that $2^2 = 4$ and $3^2 = 9$. So here's something funny: When you find the difference between the squares of the numbers 2 and 3, it ends up being the sum of the original two numbers.

$$3^2 - 2^2$$
$$= 9 - 4$$
$$= 5$$
$$= 3 + 2$$

Can you find any other numbers with a similar relationship, where the difference of the squares is the same as the sum of the two numbers? Can you find numbers that don't work?

Extra challenge: Explain why the pattern works (or doesn't work).

Task Card Book #5

—213—
Number Boxes

Draw blank boxes in the shape of any calculation, like this two-digit sum:

☐ ☐ + ☐ ☐ = ☐ ☐

Without repeating digits, can you fill in the squares to make a true statement?

What is the largest answer you can make?
The smallest?
The closest to 50?
Or to 100?

—214—
Fibonacci Growth

The Fibonacci series starts with 1, 1, 2, 3, 5… Each new number is the sum of the last two numbers before it. Write a Fibonacci series on your page and see how far you can go. What patterns do you see in the Fibonacci series?

Make up a rule for a number series of your own. Give your series a name. What questions might you ask about your number series?

—215—
Manan Shah's Twisted Cliché

Write the word "WRONG" on your page. Below it, write "WRONG" again. Draw a plus sign and a line to indicate the sum, and then write the word "RIGHT."

Can you find number values for each letter that will make the sum true? Each letter must represent the same number throughout the whole equation, which is different from all the other letters. Also, the first letter of any word cannot be zero.

What if you made the equation RIGHT + RIGHT = WRONG? Can you find values that will make that sum true?

Extra challenge: Can you find number values that will let five WRONGs add up to a RIGHT?

```
  WRONG          WRONG
+ WRONG          WRONG
-------          WRONG
  RIGHT          WRONG
               + WRONG
                 -----
                 RIGHT
```

Puzzles like this are called *alphametrics* or *cryptarithmetic*. Both of these puzzles have more than one solution.

—216—
Number Yoga

Use only the four digits __, __, __, and __ to make math expressions. [Choose any four numbers 0–9, mix or match.] You must have all four digits in each expression, with each digit appearing only once. You may use any math operation you know: +, -, ×, ÷, brackets, etc.

For example, if you choose 1, 2, 3, and 4: 1 × 2 × 3 × 4 = 24, and 34 - 12 = 22.

Can you calculate all the numbers 1–10? 11–20? What other numbers can you make? Are any numbers impossible to make with your four digits? (Or perhaps they are possible, but not with the math you've learned so far?)

Task Card Book #6

—265—
Odd Numbers Pattern

There is a cool pattern when you add consecutive odd numbers, starting with one:

$$1 + 3$$
$$1 + 3 + 5$$
$$1 + 3 + 5 + 7, \text{etc.}$$

Try it, and see what you find. What questions can you ask?

—266—
Triangle Sums

Draw a triangle with a dot at each corner. Can you label each dot with the numbers 1–3 so every side has the same sum? Why not?

Add a dot near the middle of each side and try it with the numbers 1–6, one number at each dot. Can you make the side-sums equal now?

Make a new triangle with corner dots and two dots on each side. Which numbers will you arrange this time? Can you find more than one way to do it?

—267—
Number Pyramid

Draw a stack of blocks (big enough to write in) pyramid-style, each line having one less block than the line below. Write any numbers in the bottom blocks. In each block above, write the sum of the two below it.

Try another pyramid with different numbers. What if you put a number in the top block—can you figure out which starting numbers might make that total? Can you think of other questions to ask?

Make up some number pyramid puzzles of your own. If you put numbers in some of the other blocks (not just the bottom), can you still solve the pyramid?

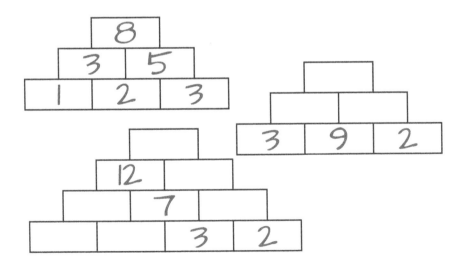

—268—
Four 4s

Use only the four digits 4, 4, 4, and 4 to make math expressions. You must have all four 4s in each expression, but no other numbers. You may use any math operation you know: +, -, ×, ÷, brackets, etc.

For example, 44 − 44 = 0, and 4 × 4 × 4 × 4 = 256.

Can you calculate all the numbers 1–10? 11–20? What other numbers can you make? Are any numbers impossible to make with four 4s? (Or perhaps they are possible, but not with the math you've learned so far?)

Mathematics is a way of thinking. It requires no tools or instruments or laboratories. It may be convenient to have a pen and paper, a ruler and a compass, but it is not essential: Archimedes managed very well with a stretch of smooth sand and a stick.

—Kathleen Ollerenshaw

Chapter 6: Geometry

At its heart, geometry is all about seeing connections and relationships. How can students break shapes apart, put them together, move them around the page, turn them, or distort them? Which properties change, and which stay the same?

Every activity has the potential to spawn hundreds of variations. Alter something in the prompt to make a fresh investigation. Tweak the size, shape, or other properties of interest. What new things can your children see in the math? What questions can they ask?

For older students, use algebra to put some teeth in the relationships they see. Give the points names. Identify the line segments. Can your students write any equations about them? Which distances are equal to other distances, or areas equal to other areas? How can they know for sure? When they add new points, lines, or circles to the diagram, what new connections do they find?

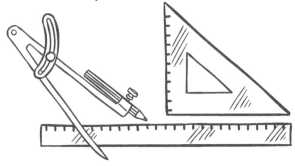

Task Card Book #1

—9—
Double the Rectangle

The length and width of a rectangle are both increased by 2 units. The new rectangle has twice the area of the original. What might be the original length and width? (There is more than one possible answer.)

—10—
Square Numbers 1

Use graph paper, lined or dotty. Draw several different squares of different sizes, counting along the grid lines to make each side the same length. When you draw a square with sides of S (any number) grid units, the area (A) will be S squared:

$$A = S^2$$

And the side is the square root of the area:

$$S = \sqrt{A}$$

Label the sides of your squares and count the square units for their areas. Make a list of the square numbers you find (the areas) and their square roots (the sides).

What do you notice about the numbers or the sizes of the squares? Can you think of any questions to ask?

—11—
Everything Is a Rectangle

Draw any *quadrilateral* (four-sided shape) on your page. How can you convert it into a rectangle with the same area? For example, can you imagine cutting off one part and pasting it in a different place?

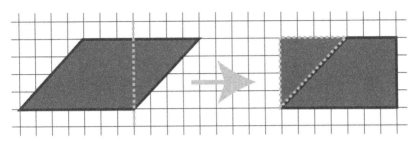

No matter how many times you cut and paste, the total area remains the same.

What if it's an unusual shape like a kite or an arrowhead? Extra challenge: Can you convert shapes that are not quadrilaterals?

—12—
Square Numbers 2

Use graph paper, lined or dotty. Can you figure out how to draw a square that's tilted on the grid? How do you know for sure it's a square?

Draw several tilted squares of different sizes. How can you find their areas, when the sides don't line up with the grid?

If you draw a tilted square on a grid, the length of its side (S) is usually not a counting number. But that square's area (A) is still the square of the side length:

$$A = S^2$$

And the side is still the square root of the area:

$$S = \sqrt{A}$$

Count out the area of your tilted squares: the whole square units plus the smaller fractional parts that go together to make units. You may need to imagine cutting and pasting pieces, as in prompt #11, to make counting easier.

Label each square's area (the square number) and its side lengths (the root).

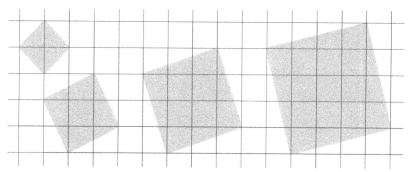

Here are four tilted squares, with areas of 2, 5, 10, and 17 grid spaces. Therefore, each side length must be the square root of that square's area.

Task Card Book #2

—61—
Perimeter Puzzle 1

A rectangle has a perimeter of _____ grid units. [Choose any number.] What might the area be?

How many different rectangles can you find with that perimeter?

What if the sides don't have to be whole unit lengths?

Extra challenge: Perimeter values less than 4 units force the use of at least one fraction or decimal side length.

—62—
Area Puzzle 1

The rectangle has an area of _____ grid squares. [Choose any number.] What might the perimeter be?

How many different rectangles can you find with that area?

What if the sides don't have to be whole unit lengths?

—63—
Hexashapes

Draw one large regular *hexagon* (six-sided shape) as precisely as you can. If you have triangle graph paper, that makes it easier. Find the midpoints of each side. Connect them to make a smaller hexagon. Draw diagonal lines inside the small hexagon so that each corner is connected with a line to every other corner. Also draw lines from the center to each corner of the large hex.

How many polygons can you find? (Triangle, rectangle, rhombus, kite, dart, etc.) How do the shapes relate to each other? For example, can you find different shapes that have the same area? Or a shape that is some fraction of another shape?

—64—
Hexangles

Draw one large regular hexagon as precisely as you can. If you have triangle graph paper, that makes it easier. Find the midpoints of each side. Connect them to make a smaller hexagon. Draw diagonal lines inside the small hexagon so that each corner is connected with a line to every other corner. Also draw lines from the center to each corner of the large hex.

How many angles can you identify without using a protractor?

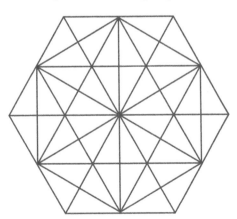

Here is what the Hexashapes and Hexangles drawing should look like before you start marking the different shapes or angles you can see.

Measure the angles to see if you were right. Measure some that you weren't sure about, to see what size they are.

How do the angles relate to each other? For example, can you find different angles that have the same size? Or an angle that is some fraction of another angle?

Task Card Book #3

—113—
Four by Four

Draw a 4 × 4 array of dots. How many different quadrilaterals can you make by connecting the dots of this array? You may want to draw several 4 × 4 arrays, so your shapes don't become a jumbled mess of lines on top of each other.

Think about squares, rhombuses, rectangles, parallelograms, trapezoids, kites, arrowheads, and irregular shapes. For which shapes can you identify the area or perimeter? What else do you notice?

—114—
Triangle Possibilities

If you keep two sides of a triangle the same length but let the other side change, you could make an infinite number of triangles. Pick lengths for the two unchanging sides. Draw some of the triangle possibilities.

Of all those infinite triangles, which one has the largest area? How do you know?

—115—
Perimeter Puzzle 2

A polygon has a perimeter of 12 grid units. [Or choose any number.] What shape might it be? If the sides of your polygon are allowed to slant, can you still make the perimeter exactly 12? What is the most interesting shape you can make?

—116—
Area Puzzle 2

A polygon has an area of 5 grid squares. [Or choose any number.] What shape might it be? If the sides of your polygon are allowed to slant, can you still make the area exactly 5? What is the most interesting shape you can make?

Task Card Book #4

—165—
Painting Blocks 1

Describe or draw how you might paint a cubic block so that each face is a single color, but no two adjacent faces share the same color. How many colors do you need?

Are there different ways you could do it? What's the largest number of colors you could use? Or the smallest?

What other rules might you make for painting a block? And what questions might you ask about those rules?

—166—
Sierpinski Triangle

Draw a large triangle. Find the midpoint of each side. Connect the midpoints. What do you notice?

In the outer three triangles (which point the same direction as the original), again find the midpoints and connect them. Repeat, and keep going, until the triangles get so small you can't draw any more.

What patterns do you see in the number of triangles in each stage of the drawing? Or in their sizes? What other questions can you ask?

—167—
Equilateral?

Is it possible to draw an equilateral triangle (with all three sides the same length) by connecting the dots in a square array? Draw it on dotty graph paper. Or if it's not possible, how close can you get?

—168—
Circles in Circles

Draw a large circle. Draw a smaller circle inside that just touches (is *tangent* to) the first. Can you draw a third circle tangent to them both? How many circles can you draw inside the big one, each tangent to at least one of the circles before?

Task Card Book #5

—217—
Pentagram

Draw a *pentagon* (five-sided shape). Draw diagonal lines inside the pentagon so that each corner is connected with a line to every other corner. The diagonals make a shape called a *pentagram*. What do you notice? What patterns do you see? Where might you add new lines to the shape?

How many questions can you ask about this shape? Can you figure out the answers to any of your questions?

—218—
Painting Blocks 2

Imagine you have two colors of paint, and each face of a cubic block must be a single color. Adjacent faces may be the same color or different from each other. How many different ways might you paint the cube?

For example, you could paint the top red and the other sides blue. Or you could paint the bottom red and the other sides blue. Are those different?

Visualizing three-dimensional shapes can be tricky. Try to imagine turning the cube in your hands. If you can rotate two cubes to make the sides match, that counts as a single way to paint it.

—219—
Don Steward's Quadrilateral Challenge

Draw several 3 × 3 arrays of dots. Can you connect the dots on those arrays to make these 16 different shapes?

- three squares
- one non-square rectangle
- two parallelograms
- one isosceles trapezoid
- two other trapezoids
- one kite
- two symmetric arrowheads
- four irregular quadrilaterals

—220—
Perimeter and Area

A shape has a perimeter of 32. [Or choose any number.] What might the area be?

Hint: You can use 3-4-5 or 6-8-10 right triangles to create shapes where the slanted hypotenuse is part of your perimeter. If you can't figure out how, look at the examples on Don Steward's blog.[†]

[†] *donsteward.blogspot.com/2015/12/area-for-fixed-perimeter.html*

Task Card Book #6

—269—
Painting Blocks 3

You have a cubic block of wood that is painted all over. Then you cut it into smaller cubes. How many cubes will you make? How many sides of each smaller cube will have paint? Will you have any cubes with no paint at all?

If you put all the cubes into a bag and pulled out one at random, what is the probability it would have paint on three sides? What other questions can you ask?

—270—
Circle Geometry

Draw a large circle. Draw its diameter. Connect both ends of the diameter to any other point on the circle. Do it again, connecting the diameter to another point, and another.

What do you notice? What questions can you ask?

—271—
The Lune of Hippocrates

Draw a circle. Then draw two diameters that cut it into quarters. Connect the ends of those diameters to make a square. With the midpoint of one side of the square as its center, draw a smaller circle that passes through the nearest corners of the square.

The moon-shape outside your original circle is the *lune*.

To find the *quadrature* of a lune or other curved shape means to match it with an equivalent rectangular shape, or to calculate its area in square units. The classic puzzle of "squaring the circle" challenged mathematicians to find the quadrature of a circle.

The quadrature of your lune exactly matches the area of another part of your drawing. Can you tell which part? Can you prove it?

—272—
The Lunes of Alhazen

Draw a large circle. Draw its diameter. Connect both ends of the diameter to any other point on the circle.

Draw a lune: With the midpoint of one side of the triangle as its center, draw a smaller circle that passes through the nearest corners of the square. Draw another lune on the other side of the triangle.

What do you notice about the lunes? Can you think of any questions to ask?

The lunes of Alhazen are a more general extension of the lune of Hippocrates.

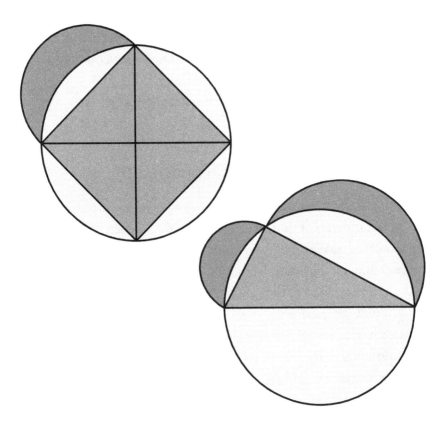

Truth and beauty are enough. I have often reminded my students that the best mathematical achievements took place when the question, "What is it for?" was not asked.

—Bhama Srinivasan

Chapter 7: Math Art

When students learn to visualize shapes, designs, and patterns, it makes them better at math. Even topics like algebra can be surprisingly visual.

Art lets children experiment with geometric shapes and symmetries. They can feel their way into math ideas through informal play. As they draw, students explore a wide range of mathematical structures and relationships.

Math doodles allow a student's mind to relax, wander, and come back to its work refreshed. And though it goes against intuition, doodling helps people remember more of what they learn.

And art is one of the most replayable of all types of journal prompts. Have you noticed how professional artists love to tinker with their creations? Even a slight change produces delightful new variations. Come back to each math art prompt and enjoy the adventure of exploring possibilities.

Task Card Book #1

—13—
Make Patterns

Create a pattern of numbers, shapes, or colors. Keep your pattern going as long as you can. Will your pattern go all the way around the page? Or begin at the center and spiral outward? Or make any other design you like.

—14—
Circle Dance

Drafting compasses can be tricky to control. Practice using a compass to draw circles on your page. Make as many as you like, in different sizes. Let some of the circles overlap each other. What do you notice about your circles?

Color your design, or fill each section with a pattern.

—15—
Circle Pattern

Draw a circle. Draw another circle the same size, with its center at any point on the first circle's circumference. At each point where the circles meet, draw another circle centered there. Continue drawing circles with their centers at each new intersection.

What do you see in this pattern?
Can you think of any questions to ask?

—16—
Create a Font

A font is a set of specially designed letters, numbers, and punctuation for use in a book, magazine, word-processing program, webpage, or

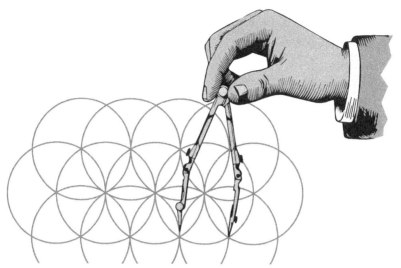

This delightful, never-ending circle pattern can be the starting point for many math art projects. Try connecting the intersection points with straight lines to make other shapes and designs.

anywhere else people may read stuff. Some fonts are very geometric and regular. Other fonts vary like freeform handwriting.

Make up your own font. You may want to use graph paper squares to keep your letters consistent. What math do you see in your font? Did you use parallel or perpendicular lines? Are there shapes that repeat in different letters?

Write your name or some other message in your font.

Task Card Book #2

—65—
Pixel Graphics

On graph paper, outline a square that is 8 × 8 grid spaces or larger. Make a black-and-white design by shading the grid squares. Each grid square represents one pixel, which must be totally filled or completely blank.

Or make a colored pixel design, with each square being one color. Extra challenge: Use only two or three colors. What pictures can you make with such a limited palette?

—66—
Connect the Dots 1

Use dotty graph paper. Draw a picture by connecting dots. Connect as many or as few of the dots as you wish. Color your design, or fill each section with a pattern.

What math do you see in your picture? Did you use symmetry? What shapes can you find?

—67—
Connect the Dots 2

Use blank paper. Draw a bunch of dots scattered around your page, numbering them as you go. Then connect the dots using an unusual rule, whatever you like.

For example, you might connect each number to its double: 1–2–4–8, then 3–6–12, and so on until you run out of numbers with doubles. Then connect the remaining numbers in order.

Color your design, or fill each section with a pattern.

—68—
Octagons

Use dotty graph paper. Connect dots to create an eight-sided shape. Are all the sides of your octagon the same length? How can you tell?

What kind of design can you make with octagons?

Will your octagon tessellate? That is, could you cover a floor with only octagon tiles? If so, draw the pattern. If not, why not?

Task Card Book #3

—117—
Squiggle Doodle Patterns

Draw a random squiggle shape on your page. Go back and forth and all around the paper without lifting your pen or pencil. Let your line cross itself as often as you like—the more crossings, the more interesting the final design. Come back to your starting point at the end to make a *closed* curve.

Fill in your doodle using two or more colors, but make sure that no two sections with the same color touch each other.

—118—
Hundred Face

Use square graph paper with a lined or dotty grid. Draw a face by coloring exactly 100 of the grid squares on your page. The squares don't have to touch each other.

In the margin, show how you counted the squares to make sure there were 100. Hint: It's easier to count accurately in groups of squares, not one by one.

[Many teachers do this puzzle with Cuisenaire rods. If you'd like to see designs from students around the world, search for the hashtags #hundredface and #hundredfaces.]

—119—
Math Comic

Draw a comic strip or comic-book-style story about math. You can use personified numbers and shapes. Or draw any characters you like but have them dealing with math in some way. Share your comic with a friend.

—120—
Sidewalk Math

Have you heard of sidewalk math? People draw math shapes, puzzles, or problems with sidewalk chalk. Look up some example pictures at sidewalkmath.com or twitter.com/hashtag/sidewalkmath.

Make up your own math design or puzzle that would work with sidewalk chalk.

Extra challenge: Find a place to draw your design outside for people to enjoy. Take a picture and share it online.

Task Card Book #4

—169—
Tricky Triangles

Use blank paper. Close your eyes and draw a triangle. Open your eyes to see how you did. Try again. Try other shapes.

—170—
Symmetry Design

Use graph paper, lined or dotty. Draw a line on the page, or two lines that cross at a right angle. Create a design that uses the line(s) as a mirror reflection. Color your design, or fill each section with a pattern.

—171—
Symmetry Puzzles

Use graph paper, lined or dotty. Draw a line on the page, or two lines that cross at a right angle. Draw only part of a design that uses the line(s) as a mirror reflection. Trade with a friend and try to finish each other's picture.

Color your designs, or fill each section with a pattern.

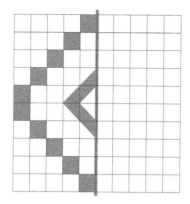
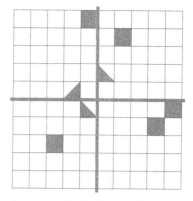

Symmetry puzzles can be relatively easy or devilishly tough, depending on how much of the design you draw. Both of these puzzles create the same final picture. Can you see how?

—172—
Low Poly Art

Low poly art is a graphic design style used in video games and animation. And it's a cool way to create a modern-looking, minimalist picture. Draw a picture that is all polygons. But use the simplest polygons you can—mostly triangles, with only a few quadrilaterals where necessary.

Anywhere two polygons meet, they should share a complete side with its vertexes. The polygons meet corner-to-corner, like a fishing-net mesh, not side-to-corner like a wall of bricks. Color your design, if desired.

Extra challenge: Mix your tints and shades so any two polygons that share a side have different colors.

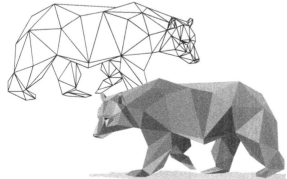

The shapes in low poly art fit together, sharing their sides and vertices.

Task Card Book #5

—221—
Make a Maze

Use dotty graph paper. Connect the dots on your page to draw a maze or labyrinth with horizontal and vertical passages. Include a few diagonal passages, too, if you like. Make your design as complex as you like, but be careful to leave at least one path through your maze.

—222—
Rotational Symmetry

Use graph paper, lined or dotty. Draw a large square. Draw lines to divide the square into fourths. Create a design in one section. Then repeat it in each of the other sections, giving your design a quarter-turn each time, so it rotates around the center of the square.

If you turn your square like a spinner, will the patterns match up? These directions make a design with four-fold rotational symmetry, also called "symmetry of order 4." Use triangle graph paper to create a design with three-fold or six-fold rotational symmetry.

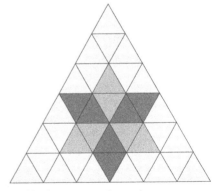

Rotational symmetry is named by how many times the design matches itself as you spin it. 4-fold symmetry matches in four positions, 3-fold symmetry in three positions.

—223—
Tessellations

Use dotty graph paper. Tessellations are patterns that repeat the same shape all over the page, like checkerboard squares or hexagons in a honeycomb.

Can you connect grid dots to make a shape other than a square that will tessellate? Or more than one shape that work together to make a tessellation? Fill the page with your shape(s), and then color your design.

Your tessellation shapes may turn in different directions, like the rhombuses (diamond shapes) in this Roman mosaic.

—224—
Nrich Tablecloths

Draw several 5 × 5 square grids. Imagine that each grid is a tablecloth of quilted squares, 25 squares per tablecloth. Color the grids so that every square is a single color, like pixel art.

Use as many colors as you like, but the design for each tablecloth grid must have some symmetry. You can use lines of symmetry (1, 2, or 4), rotational symmetry (order 2 or 4), or both. How many different tablecloth designs can you find?

Extra challenge: Investigate different sizes of square grids. See examples at the Nrich "Attractive Tablecloths" investigation.[†]

[†] nrich.maths.org/900

Task Card Book #6

—273—
Artist's Choice

Outline any pattern or design you like. Can you color your design using only two colors, so that no two sections with the same color touch each other? If so, can you make a new design that is impossible for only two colors? Or if your first design wasn't possible, can you make a new one that is?

—274—
Harder Than It Looks

Using a ruler looks easy, but the silly things like to slide around while you're trying to draw with them. Practice using a ruler to draw straight lines with no slipping or finger-bumps. Make a pattern or design entirely from straight lines. Color your design, or fill each section with a pattern.

Challenge for older students: Draw a cardioid. Start with a big circle. Add a lot of dots (30–60) around its circumference, and number them in order. Draw straight lines to connect each dot number to its double: 1 connects to 2, 2 to 4, and so on. Imagine that the numbers keep going around past the first dot, so 20 connects to 40, even if you don't have that many dots. Can you see how the lines form a heart shape?

Extra challenge: What happens if you connect each number to its triple, or quadruple?

A *cardioid* is a rounded-heart-shaped curve.

—275—
Mandala

Start with a point in the center of your page. Draw a circular, symmetric pattern or design all the way around the point. Draw another pattern to make a second-layer design. Then another. Continue adding layers around your mandala, keeping the overall rotational symmetry. Some layers may be smoothly circular while others zigzag like flower petals. When you like your design, stop and color it.

—276—
Warped Chessboard

Draw an irregular grid of lines that cross each other. The lines do not have to be vertical and horizontal, nor evenly spaced, nor parallel, nor straight. Decorate the shapes formed by your lines with colors or patterns that alternate, like the dark and light squares on a chessboard.

What do you notice about your design? Can you think of any questions to ask?

Writing organizes and clarifies our thoughts. Writing is how we think our way into a subject and make it our own. Writing enables us to find out what we know—and what we don't know—about whatever we're trying to learn.
—WILLIAM ZINSSER

Chapter 8: Writing

WRITING HELPS STUDENTS STRETCH THEIR thinking and make sense of new ideas.

When students wrestle their thoughts into shape and create explanations, they do the same sort of work that mathematicians do every day. It's difficult for children (or anyone) to capture a thought and cage it in words. But it's great practice for life.

Students may supplement their writing with illustrations. Sketch drawings can be a wonderful aid to mathematical thinking.

For the poetry prompts, students should aim for evocative descriptions, vivid verbs, and playful words. If your child can't think of where to start a poem, try brainstorming a list of sensory details.

You may reuse writing prompts as often as you like. Change the question, if you wish—but even when the prompt remains the same, the students have changed since the last time they wrote about it. Today is a new day, so they are seeing with fresh eyes and thinking different thoughts.

Task Card Book #1

—17—
Noticing

When was the last time you found something mathematical in actual life? Describe what you saw and tell how it relates to math. Or if you can't think of anything, then look at the room around you and describe some math you can see. Think about numbers, shapes, lines, curves, and patterns.

—18—
Silly Definitions

Flip ahead in your math book. Find three vocabulary words you don't know. Think up serious or wacky definitions for them. If you wrote the math dictionary, what would these words mean?

—19—
Six-Word Stories

Summarize a math concept or lesson topic in six words. Write it like a riddle: "Three lines. Three angles. Closed shape." Or name the topic in your story: "Three lines. Three angles. One triangle."

—20—
Math Poetry: The Square

Write a poem in which every line has the same number of words as the entire poem has lines. Try to use sensory details and vivid verbs. Your poem does not have to rhyme, but it can if you wish.

For example, you could write six lines with six words in each line. Like this…

Exponential Adventure

Once when famine struck the land—
No rain, no hope of harvest—
The youngest princess launched her quest.
She sought the fabled magic chessboard
That double, double, double, doubles rice.
Enough for all her starving people.

Extra challenge: Try a longer cubic poem, with the same number of stanzas as it has lines per stanza and words per line. Or a hypercube epic with sections, stanzas per section, lines per stanza, and words per line.

Task Card Book #2

—69—
Growth Mindset

Have you heard that your brain keeps growing the more you use it? And that mistakes help you learn even more than when you get things right? How do these scientific discoveries affect your attitude toward math?

—70—
Museum of Math

Did you know there is a National Museum of Mathematics in New York? You can learn more about it at the website momath.org.

Imagine you are the curator designing a new exhibit for the Museum of Math. What will you put in your display? Will you make it interactive? How?

—71—
Math Poetry: Haiku or Senryū

Count syllables to make a non-rhyming mathematical poem. Haiku have three phrases (lines) with seventeen syllables in a five-seven-five pattern. Usually, the theme of the poem is expressed by a key word at the end of one phrase.

Senryū is haiku's more cynical cousin, exposing human foibles and often laced with dark humor. For example:

MULTIPLICATION
How much ice cream in
Five triple-dip fudge sundaes?
Getting fat with math.

—72—
Frayer Model

Choose a math vocabulary word. Define it with a Frayer model. First, write the word in the middle of your page and circle it. Then draw lines to divide the remaining page into quarters with labels: definition, characteristics (or illustration), examples, and non-examples. Fill in each section.

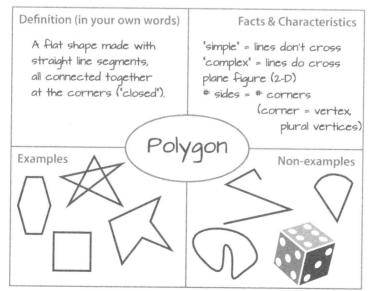

Task Card Book #3

—121—
Math Poetry: Word Equations

Write a non-rhyming math poem that has the format of a calculation, but with words instead of numbers. Try to think of clear, vivid words and phrases that create a picture in your reader's mind.

Author Betsy Franco called word equation poems *Mathematickles*. They are fun and easy for all ages. Here are some examples:

> *Emerald-green grass*
> *+ Tiny drops of sunshine*
> *= Springtime dandelions.*

Or…

> *Freshly baked cookies ÷ Children*
> *= Delight R Crumbs.*

[Where "R" signals the remainder.]

—122—
Automathography

Write your math autobiography. Tell the story of your mathematical education. What is your earliest memory of math? Or tell one of your best (or worst) experiences with math.

What parts of math have you enjoyed or found interesting? Was math ever your favorite subject? What made you like it? If not, what didn't you like?

Can you remember a time when someone helped you understand a math topic? Tell about it. If not, then tell how a helpful mentor could have made a difference to you.

Was there a time when your opinion about math changed (for better or worse)? Explain.

—123—
Math Comparisons

How are _____ and _____ alike? How are they different? What questions can you think of to ask about them?

Hint: Pick any two numbers, shapes, or math topics the student has learned. The question is often more fun when the topics are not obviously related.

—124—
Snail Mail

Write a letter to a mathematician (living or dead) or to mathematics itself.

Optional challenge: If you write to a living mathematician, copy the letter in your best handwriting and mail it.

Task Card Book #4

—173—
Versatile Tools

How many different, unusual ways can you think of to use a ruler? (Or a compass, protractor, dice, etc.)

—174—
Something Interesting

Write about an interesting thing you've learned in math. It doesn't have to be a school topic—think about infinity, or fractals, or tessellations, or any other wild math concept. What do you know about it? What questions can you think of to ask?

—175—
Math Poetry: The Fib

Write a non-rhyming poem in which the first two lines have only one short syllable each. After that, every line has the same number of syllables as the two preceding lines combined. That is, the number of syllables per line form a Fibonacci sequence. For example:

The Story Problem

Read.
Then
Again,
Read it through.
What is it asking?
How can you use the things you know
To figure out the mystery of the great unknown?

—176—
Dear Younger Me

What do you wish someone had told you about math? Or about solving problems, or learning in general? Write a letter to your younger self, explaining what you wish you'd known. Or write to a sibling, cousin, or other child.

Task Card Book #5

—225—
Familiar Math

Since you got up this morning, what math have you used? If your first answer is "none" then think again. Have you used any numbers or shapes? Have you seen any patterns? Have you thought about how big or small or cold or hot something is? Or any other measurement?

—226—
Questions

Some mathematicians say being able to ask good questions is even more important than answering them. Write down a list of questions about math. How many different things can you ask?

—227—
Math Pietry

Write a math poem (any form) about the number *pi*, which is how many times you have to walk across a circle to equal one time walking around its edge.

Pi = circumference ÷ diameter

Or write a pi mnemonic, where each word has as many letters as that digit of pi: 3.1415926…

Or try pi-ku, a haiku-like poem where each line has syllables matching one of the digits of pi. For a six-line pi-ku, arrange your thoughts in a 3–1–4–1–5–9 pattern. For example:

Journaling Pi-ku

Math makers
Forge
Thoughts into words
To
Create something new,
Exploring numbers, shapes, and patterns.

—228—
Let's Get Meta

Metacognition means thinking about your own thinking. Look back at an earlier entry in your math journal. Did it communicate what you meant to say? Has your thinking changed since then? What would you say differently if you wrote about that topic today?

Task Card Book #6

—277—
Words Help Us Think

Make a math "word wall" on your journaling page. Write all the math vocabulary words you know. Decorate them with frames, or write them crossword-style or in different directions. Or make it fancy any way you like.

Optional: Include definitions or pictures explaining the hardest-to-remember words.

—278—
A World with No Math

Imagine you went through a magic portal to a place that doesn't use math. No numbers, shapes, measurements, or patterns. No money or clocks. Describe what it would be like to live there.

—279—
Math Poetry: Acrostic

Choose a math word. Write it in capital letters vertically down your page, one letter under another. On each line, write a word or short phrase that begins with the letter and is related to the main word.

Alternative form: Each line contains that letter somewhere within it, not necessarily at the beginning of the word or phrase. Try to line up the letters so the word is easy to read. For example:

>matheMatics
>tantAlizing
>creaTive
>thougHt-provoking

—280—
Solving Problems

Metacognition means thinking about your own thinking. Write about the way you solve math problems. Choose a specific problem and describe your thinking as you work on it. Or write about mathematical thinking and problem-solving in general.

What would mathematics have amounted to without the imagination of its devotees—its giants and their followers? There never was a discovery made without the urge of imagination—of imagination which broke the roadway through the forest in order that cold logic might follow.

—David Eugene Smith

Chapter 9: Freewrites

You can spark creative thought by removing any need to worry about spelling or punctuation rules. During a freewriting session, students should write fast and raw until they reach the end of the page.

If students can't think of what to write, they might consider how the 5W1H questions apply to their prompt: who, what, where, when, why, how? Or they can reword their previous sentence, or look for a way to add extra details. It's always valuable to rethink and revise our writing to make it better express the ideas in our heads.

If all else fails, students can keep writing anything that comes into their minds, even if it seems to have nothing to do with math. They may be surprised to find mathematical ideas pop up in the most unexpected places.

Freewriting prompts may be reused, especially if you change one or two words to make them new. Most of the questions are general enough to spawn entire books, so there will always be ideas the student didn't have time to think about before.

Or let students propose their own topics. Make a long list of prompt sentences and cut the paper into strips, then crumple the strips into a jar. When it's time to write, they can pull out a prompt and let the pencil run with it.

If children have trouble filling a whole page, let them write to a timer instead. Set it for five minutes or however long fits their energy level.

Task Card Book #1

—21—
Learning

The best way to learn math is…

—22—
Stuck

When I get stuck on a math problem…

—23—
It's Easy

The easiest mistake to make in math is…

—24—
Memory

In math, it's always important to remember…

Task Card Book #2

—73—
Games

What kind of games do you prefer, chance or strategy? Explain why.

—74—
Analogy

Math is like…

—75—
The Adventure of Learning

At first, I thought _____. But then I discovered _____. Now I wonder _____.

—76—
Brain Dump

Tell what you know about _____. Or, what are the most important things to understand about _____? [Choose any math topic, especially something broad like "addition" or "percents."]

Task Card Book #3

—125—
Feelings

Math makes me feel _____ because…

—126—
Fun

When I hear someone say that math is fun…

—127—
Always Learning

I wish I knew more about…

—128—
Mistakes

When I make a mistake in math...

Task Card Book #4

—177—
Wondering

Is math important? Why, or why not?

—178—
Frustration

I always get confused/frustrated when...

—179—
A Good Teacher

What are some of the qualities of a good math teacher? What does a good teacher do, or not do? Explain.

—180—
A Good Student

What are some of the qualities of a good math student? What does a good student do, or not do? Explain.

Task Card Book #5

—229—
Math Debate

Is math easy or hard? Explain.

—230—
School

Should everyone have to study math? Why, or why not?

—231—
Keep It Simple, Sometimes

Is it important to simplify fractions? Why, or why not?

—232—
Maps

Imagine (or look at) a map of your city, state, or country. Is a map math? Why, or why not?

Task Card Book #6

—281—
Money

Is money math? Why, or why not?

—282—
Success

One thing I'm proud of in math is…

—283—
Do-Over

If I could do anything in math over again, I would…

—284—
The Science of Patterns

Some people say mathematics is "the science of patterns." Describe some patterns you've noticed in math.

The mathematical question is, "Why?" It's always Why. And the only way we know how to answer such questions is to come up, from scratch, with these narrative arguments—these elegant reason-poems—that explain it.
—Paul Lockhart

Chapter 10: Explanations

MATH JOURNAL EXPLANATIONS AVOID THE formality that turns so many students away from geometry proofs. These informal "reason-poems" drive at the heart of a student's understanding. How did they figure this out? Why does their method work? Is the pattern they found real or just a temporary coincidence? How do they know?

When you run out of creative journaling ideas, you can always go back to the basic mathematical question: "Why?"

For older students, challenge them to explain a concept so that a kindergarten student or second grader could understand. That's more difficult than it sounds, but the attempt forces students to clarify their own ideas about the topic.

Task Card Book #1

—25—
Mystery Coins

I have eight coins in my pocket. How much money might I have? Is there any amount that's impossible? Why? Explain your answers.

—26—
Colored Paper and Metal Disks

What is money? Could you explain it to an alien from a planet that doesn't have money?

—27—
Math Translation

Find an equation in your math textbook. Translate it into words with no symbols at all.

Extra challenge: Read your translation to a friend. Can they write the original equation from your description?

—28—
Explain a Mistake

Describe a mistake you made in math, or a problem you missed on a quiz or test. What went wrong? How will you avoid this error the next time? Do you understand the problem now, or is there something more you need to learn about it?

Task Card Book #2

—77—
Make It Visual

Draw a picture or diagram to explain something you've learned in math. Label it so anyone could look at your picture and know what you mean.

—78—
Explain a Problem 1

Go to expii.com/solve and find a math problem you like. Copy it in your journal. Or copy a problem from your math book. Explain how to solve it. Try to make your explanation clear enough for a younger sibling, cousin, or friend to understand.

Extra challenge: Can you show more than one way to figure it out?

—79—
Explain a Puzzle

A stick has two ends. If you cut off one end, how many ends will the stick have left?

A square has four corners. If you cut off one corner, how many corners will the remaining figure have?

Is subtraction broken?

—80—
Explain a Problem 2

Copy a story problem from your math book, but don't include the numbers. Can you explain how to solve it without using numbers?

For example: "Joseph knows the price of a box of candy and the price of a certain book. How can he figure out how much money he will have left after buying them both?"

If you enjoy the challenge of solving problems without numbers, you might like Farrar Williams's book *Numberless Math Problems: A Modern Update of S.Y. Gillian's Classic Problems Without Figures*.

Task Card Book #3

—129—
Strategic Thinking

Play a game. Describe your strategy for winning. How do you plan your moves? How do you figure out responses to your opponent's moves?

—130—
Explain How 1

Think of something you have learned to do in math. Write an explanation simple enough that a younger sibling, cousin, or friend could understand how to do it. Add pictures, charts, or diagrams if they help to make your meaning clear.

—131—
Explain How 2

What is the hardest thing you've learned in math? Explain how it works.

—132—
Plan a Math Mini-Zine

Choose a math topic to explain. Draw eight rectangles to make a storyboard. In one rectangle, write the title. Then sketch out your explanation in the remaining blocks.

When you like your plan, fold a piece of paper and create your math mini-zine.

You can find example math zines at the Public Math website. And if you need help with the folding instructions, see Paula Beardell Krieg's directions online.†

---
† *public-math.org/zines*
bookzoompa.wordpress.com/2009/11/30/how-to-make-an-origami-pamphlet

The most confusing step is knowing where to cut. When your paper zigzags like a W, cut the middle hump in half.

Task Card Book #4

—181—
The Math of Math Tests

Would you rather take a math test that has 10 questions worth 10 points each?
　Or one with 20 questions worth 5 points each?
　Or one big problem worth 100 points, or 25 questions worth 4 oints each?
　Or some other arrangement? Explain why.

—182—
Remembering

What is something you want to remember so you can use it in the future? Explain why this is important.

―183―
The Greatest What?

Who is the greatest, and can you prove it? Choose your own category of greatness. Use math to explain why that person should win the prize trophy for "Greatest _____."

―184―
How Do You Know?

Can you tell if a math answer is correct without looking in the solutions book? How? Could you do it without a calculator? Does it depend on what kind of math you're doing?

Task Card Book #5

―233―
Applied Math

Explain how math is used in your favorite sport or hobby.

―234―
Surprise!

Write down any three-digit number ABC. Repeat the digits to form a six-digit number ABCABC.

Divide the number by 13, ignoring any remainder (the digits after the decimal point, if you use a calculator).

Divide that answer by 11, ignoring any remainder. Finally, divide by 7.

Can you explain what happened?

—235—
Payday Puzzle

You just got a new job. Your boss asks how you want to be paid:
$100 for every day that you work.
Or $10 on your first day, $20 on the second day, $30 on the third day, etc.
Or $1 on your first day, $2 on the second day, $4 on the third day, $8 on the fourth day, etc.
Which would you choose, and why?

—236—
Mental Math

Think of a calculation that you can do in your head. Explain how you figure it out. Example: "$7 \times 25 = 175$ because I think of money: 4 quarters is a dollar, and 3 more are 75 cents. So the total is 175 cents."

Task Card Book #6

—285—
The Best Team

Jacey's team won 8 out of 10 games. Sonya's team won 14 out of 18 games. If the teams played each other, who do you think would win? Explain your answer.

—286—
Explain a Math Trick

Do you have any "tricks" or mnemonics that help you remember things in math? Explain how your trick works. Is your trick always true, or does it only work in special situations?

—287—
True Grit

Think of a math problem you struggled to understand and solve. Describe how you overcame it. Looking back, do you think this struggle was valuable? If so, in what way? Or if not, then tell what would have helped you with the problem.

—288—
Beauty

Many mathematicians have written about how they find math beautiful and creative. What do you think they mean? Is mathematical beauty similar to or different from other kinds of beauty?

It is not knowledge, but the act of learning, not possession but the act of getting there, which grants the greatest enjoyment. When I have clarified and exhausted a subject, then I turn away from it, in order to go into darkness again. I imagine the world conqueror must feel thus, who, after one kingdom is scarcely conquered, stretches out his arms for others.

—KARL FRIEDRICH GAUSS

Chapter 11: Research Reports

WHERE DID MATH COME FROM? Who thought up the rules? How are the ideas and methods we study used in actual life?

Research prompts help students view math as a human endeavor, something that grew bit by bit as people in many countries across many centuries struggled to understand their world.

William Berlinghoff and Fernando Gouvêa, the authors of *Math Through the Ages: A Gentle History for Teachers and Others*, explain:

> *"It is all too common for students to experience school mathematics as a random collection of unrelated bits of information. But that is not how mathematics actually gets created. People do things for a reason, and their work typically builds on previous work in a vast cross-generational collaboration.*
>
> *"Mathematics, after all, is a cultural product. It is created by people in a particular time and place, and it is often affected by that context. Knowing more about this helps us understand how mathematics fits in with other human activities."*

While most research journaling is informal, older students may wish to venture into writing more formal reports. Here they can practice essay skills like formulating a *thesis statement* (a concise summary of the student's main idea) and organizing facts and information to prove their point.

Be sure to allow plenty of time for students to respond to research-based journaling prompts.

Task Card Book #1

—29—
Ancient Numbers

Look up how to write numbers with Roman or Mayan numerals. Can you write your birth year in that style? How would you do math calculations with numbers like these? Make an infographic using pictures and words to explain the numbers.

—30—
Mini-Biography 1

Write about a Black mathematician who is NOT Benjamin Banneker. Or write about a Latino or Indigenous mathematician.

Look at mathigon.org/timeline or arbitrarilyclose.com/mathematician-project or mathshistory.st-andrews.ac.uk/biographies for ideas.

—31—
Career Math

What career are you considering? Write about the ways you will use math in that job. If you're not sure, interview someone with that career.

—32—
Living the Dream

If you won $1 million, what would you do with it? Would you have to pay taxes on it? What would you buy? Would you save any of the money, or use it for college, or give some away?
[For younger children, use $100 or $1,000.]

Task Card Book #2

—81—
Hieroglyphic Numerals

Look up how to write numbers with Egyptian hieroglyphic symbols. Can you write your birth year Egyptian-style? How would you do math calculations with numbers like these?

Make an infographic using pictures and words to explain hieroglyphic numbers.

—82—
Mini-Biography 2

Write about a female mathematician who is NOT Ada Lovelace.

Look at mathigon.org/timeline or arbitrarilyclose.com/mathematician-project or mathshistory.st-andrews.ac.uk/biographies for ideas.

—83—
Math Report

Read a math article. Look up something on mathigon.org or nrich. maths.org or mathsisfun.com. browse the old mathmunch.org blog or search for Martin Gardner's classic *Scientific American* columns.

Write about what you learned. What questions can you ask?

—84—
U.S. Presidents

Who do you think lived longer, presidents from long ago or from more recent times? Why?
[Not from the United States? Use the rulers of your own country.]
Now check the data, to see if your prediction was correct. Make a list of all the dead presidents and the age at which they died, then split them into two groups for comparison. What do you see? What questions can you ask?
Hint: This is an excellent project for a stem-and-leaf or box-and-whiskers graph to compare the older and modern groups. See Don Steward's blog for examples.[†]

Task Card Book #3

—133—
Binary Numbers

Binary numbers have been described as "math so simple a computer can do it." Look up how to write binary numbers, if you don't already know.
Make an addition and multiplication table in binary. What patterns do you see? What questions can you ask?
Make an infographic using pictures and words to explain binary numbers.

—134—
Mini-Biography 3

Write about a mathematician who is still alive.
Look at mathigon.org/timeline or arbitrarilyclose.com/mathematician-project or mathshistory.st-andrews.ac.uk/biographies for ideas.

---

† donsteward.blogspot.com/2012/01/comparing-two-data-sets.html

—135—
Math History

What is the story of math? How do you think people began learning about math, long ages ago? How did their understanding grow over the centuries?

—136—
Earth's Belt

Suppose you could tie a rope belt around the earth's equator. How long would it be?

If you added one meter to the length of the belt, how high would it be above ground? Could an ant crawl under it? Or a mouse?

Now suppose you wanted a new rope that's one meter high all the way around. How much longer than the original belt would that have to be?

Hint: The earth is not quite a sphere because it bulges a bit around the equator and is flatter at the poles. NASA reports that the radius of Earth at the equator is 6,378.137 kilometers.

Task Card Book #4

—185—
Captain's Log

Keep a math log for a week. Write down all the math you do each day, not just schoolwork. Try to write at least one sentence every day. Did you learn anything new? Did anything surprise you?

—186—
Mini-Biography 4

Write about a mathematician who is NOT from Europe or North America.

Look at mathigon.org/timeline or arbitrarilyclose.com/mathematician-project or mathshistory.st-andrews.ac.uk/biographies for ideas.

—187—
Math News

Choose an event or discovery from math history. Write a news report about it.

—188—
Estimation

How many children live in your city? [Or state, province, country, etc.] First, take a guess. What number do you think is too big? Too small?

Then look online for data that will help you make a better estimate. Explain your reasoning.

Task Card Book #5

—237—
Math Interview

Interview an adult about how they use math in their daily life. Write about what they say.

—238—
Mini-Biography 5

Write about a mathematician whom you think most people have never heard of.

Look at mathigon.org/timeline or arbitrarilyclose.com/mathematician-project or mathshistory.st-andrews.ac.uk/biographies for ideas.

—239—
In the News

Read a news article. Write about how it connects to math. Are there parts of the story that could be counted, measured, or graphed? Are there data that indicate a trend?

Examples: slowrevealgraphs.com and nytimes.com/column/whats-going-on-in-this-graph.

—240—
Writing Math

Choose any math symbol (+, =, π, etc.) and find out when it was first used. Did people use other symbols before that? Or did they have another way to communicate the mathematical idea? What else can you discover?

Task Card Book #6

—289—
Mathematicians

What is a mathematician? What does one do? Why would anyone want to study math for their whole life?

—290—
Mini-Biography 6

Write about any mathematician you like.

Look at mathigon.org/timeline or arbitrarilyclose.com/mathematician-project or mathshistory.st-andrews.ac.uk/biographies for ideas.

—291—
All Tied Up

Look up how to make Celtic knot designs. Here is one post that might help, and it includes a link to special dot paper: mathhombre.blogspot.com/2011/03/knot-fun.html.

Or try this approach, which works with any paper: entrelacs.net/-celtic-knotwork-the-ultimate.

Create several Celtic knot patterns, using whichever method you prefer. What math do you see in your patterns? What questions can you ask?

—292—
Persuasive Evidence

Write a persuasive essay. Think about something that you believe should change, and try to convince a reader to agree. Find some real-world data to back up your opinion. Explain how the data support your argument.

Math is not neutral, nor objective, and we all bring our own perspectives to it. Data and graphing is a beautiful place to highlight this idea and value the understandings that students bring to the texts we present them.
—KASSIA OMOHUNDRO WEDEKIND AND CHRISTY HERMANN THOMPSON

Chapter 12: Measurement and Data

MEASUREMENT IS OUR WAY OF connecting numbers to the things we find in the world, in daily life. Those numbers become data that students can examine, compare, and reason about.

Some measurements are clear and easy to determine, such as the length of a stick or the weight of a bunch of bananas. But other measurements are fuzzy and open to debate. For example, how can anyone measure the value of a painting or the intelligence of a puppy?

The prompts in this book give students a chance to collect and examine a variety of measurements and to practice different ways of representing data with charts or graphs.

Older students may want to examine how data shape the way people understand their society. Two websites to explore: slowrevealgraphs.com and nytimes.com/column/whats-going-on-in-this-graph.

Task Card Book #1

—33—
Math About Me

Create a page with data about yourself. For example, you might include your height, weight, age, number of people or pets in your family, your birthday, favorite number or shape, history (5 years ago, I…) or future (in 3 years, I will…).

Extra challenge: Write the facts math-rebel-style, using an expression in place of each number. "I have 9/4 dogs."

—34—
Comparison Puzzles

A melon weighs as much as 5 premium apples, or as much as 15 plums. How do plums and apples compare?

Make up a measurement-comparison question of your own.

—35—
Fraction Wall

Near the top of your page, draw a long, thin rectangle. Below that, draw an identical rectangle, and then divide it exactly in half. How can you measure where to draw the dividing line?

Below that, draw more rectangles and divide them exactly into thirds, fourths, etc. Which fractions are the easiest/hardest to measure?

Color your fraction wall and label the parts.

(For an example, see the picture on page 120.)

Hint: One of the easiest ways to measure fractions is to cut a strip of scratch paper and fold it into equal parts.

A fraction wall can include as many or as few fractions as you like.
Feel free to skip the more difficult fractions, like fifths and sevenths.

—36—
That's Mean 1

The *mean* (average) of four numbers is ___. [Choose any value.] What might the numbers be? What else might they be?

Extra challenge: For each set of numbers you find, also figure out the median, mode, and range.

Task Card Book #2

—85—
Sorting Collections

Collect a bunch of small items: buttons, Lego blocks, coins, etc. Dump them on the table. What math can you see in your collection? What categories or attributes might you use to sort the items?

Record your observations. What questions can you ask? How can you describe the collection with math? Would a chart or graph be useful?

—86—
Collecting Data

Choose something you are curious about that can be measured. For example: the outdoor temperature or rainfall, or how many jumping jacks you can do in a row, or how much time you spend on social media.

Measure it every day for at least a week. Write a list of things you notice about the numbers that you find. Make a chart or graph to visualize your data. Remember to label your chart or graph, so people can tell exactly what you measured.

Older students: Try several different styles of graph. Which one do you like best for this type of information?

—87—
John Golden's Fraction Square

Outline a large square on your page. Draw lines to divide it into four or more fractions, all different from each other. How do you know the size of each piece? Do these fractions really add up to one complete square?

"There's something about this assignment that really reveals some strange corners of learners' fraction understanding." —JOHN GOLDEN

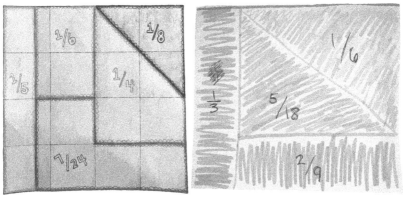

Two students' not-quite-successful attempts at the Fraction Square puzzle. Both had trouble naming the fraction pieces. Can you identify the errors?

—88—
That's Mean 2

The mean (average) of five numbers is ___. [Choose any value.] When you add another number to the data set, the mean goes up [or down] by ___. What was the sixth number?

Task Card Book #3

—137—
Clock Puzzle

The hour hand and minute hand make a right angle. What time might it be? What other questions can you ask?

Extra challenge: Older children may want to take into account that the hour hand moves, so 12:15 is not exactly a right angle. How will you measure that movement?

—138—
Interior Decorating

Measure a room in your house. Draw a scale model on graph paper. On another sheet of graph paper, make scale models of the furniture and cut them out. Using your models, decide how you'd like to rearrange the room.

Hint: Try cutting the furniture from sticky note pads, using the same scale as the graph paper.

—139—
Comparing Data

Make an obstacle course. Time yourself and your friends as you go through it. Do it again, until each person has run the course at least three times. Write a list of things you notice about your data.

Make a chart or graph to visualize your data. Remember to label your chart or graph, so people can tell exactly what you measured.

Older students: Try several different styles of graph. Which one do you like best for this type of information?

—140—
That's Mean 3

Make a list of four or five algebraic expressions of the form $Ax + B$. Make sure that A and B are different in each expression. You may use any variable (it doesn't have to be x), but keep the variable the same in every expression.

Now, find the mean (average) of your expressions. Check your answer by substituting a number (like $x = 4$) for your variable. Is the mean you found true for that number?

Task Card Book #4

—189—
Candy Graph

Get a bag of small candies that come in different colors. Sort the candies according to color, and make a graph of your results. List at least five things you notice.

Older students: Check several bags of candy. How do they vary?

Try different styles of graph to represent your data. Which one do you like best for this type of information?

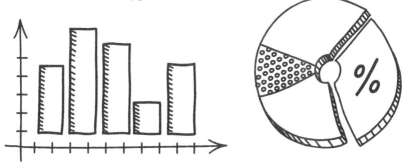

MEASUREMENT AND DATA ♦ 123

—190—
Units Count

How old are you? Does it matter what units you use to measure? Can you figure out how many days old you are? Are you older than 10! seconds?

The exclamation point after a number is the math symbol for *factorial*. Factorial counts the *permutations* of that many items, which means the number of ways you can arrange them in order in a row.

You calculate the factorial of a number by repeated multiplication of all the numbers down to one. For example, ten factorial would be:

$$10! = 10 \times 9 \times 8 \times 7 \times \ldots \times 1$$

—191—
Whatever the Weather

Find a source of weather data for your area that includes both current and historical measurements. Write a list of things you notice about the numbers that you find.

Make a chart or graph to visualize your data. Remember to label your chart or graph, so people can tell exactly what you measured.

Older students: Try several different styles of graph. Which one do you like best for this type of information?

—192—
That's Mean 4

Create a data set with 5–10 numbers of your choice. Find the mean of your set. Now split the set into two or more groups, not necessarily the same size, and find the mean of each group. Does the mean of these means equal the mean of the whole set?

What if you tried different groups, or a different original set?

Task Card Book #5

—241—
Number Race

Write the numbers 1–12 down the side of your page, each on its own line. Outline a row of four boxes next to each number (or more boxes for a longer race). Roll two dice, and shade in a box for the sum. The number that fills its boxes first wins the race.

Try another race. Does the same number always win? Which number can never win, and why? What other questions can you ask?

—242—
Benchmarks

Make a list of benchmark measurements. *Benchmarks* are numbers you can use to help you estimate other quantities. What do you know that is about 1 centimeter long? 10 centimeters? 1 foot? 1 pound or 1 kilogram? 1 ounce or 1 gram? etc.

—243—
Sets with Rules

Choose a starting number and a two-step rule for calculating the next number. For example, your rule might be "×4, then −7."

Generate a data set of five numbers: your starting number and three generated by the rule. Find the mean, median, and range of your set. Is there a mode?

—244—
People Count

Find a source of population, age, income, education, or other data for your city, state, province, or country. Try to find both current and historical measurements. Write a list of things you notice about the numbers that you find.

Make a chart or graph to visualize your data. Remember to label your chart or graph, so people can tell exactly what you measured.

Older students: Try several different styles of graph. Which one do you like best for this type of information?

Task Card Book #6

—293—
How Many?

Put a bunch of something in a clear glass or jar, without counting. How many do you think there are? Write down a number you think is too high, and one that's too low. How do you know?

Then write down your best guess. Take out a sample of the objects and count them. Do you want to revise your estimate?

—294—
Math Eyes

Imagine you have special glasses that let you see everything through the lens of math. Choose an item from your room or something in your house. Don't tell what it is, but describe it using as much math as you can. Think about measurements, shape, position, designs or patterns, motion, etc.

Extra challenge: Read your description to someone. Can they identify the object?

—295—
Budget Busters

How much money does the government of your city, state, province, or country spend in an average year? How has this amount changed over time?

If you had that much money in ordinary bills or coins, how big would your piggy bank need to be? Or how big of a box would you need to store all the cash? Would it fit in your home or school? What other questions can you ask?

Make a chart or graph to visualize the data you find. Remember to label your chart or graph, so people can tell exactly what you measured.

Older students: Try several different styles of graph. Which one do you like best for this type of information?

—296—
Equal Averages

For a certain data set of four numbers, the mean = median = mode = range. What might the four numbers be? What else might they be? Can you find a general pattern?

The experience of mathematics should be profound and beautiful. Too much of the regular K–12 mathematics experience is trite and true. Children deserve tough, beautiful puzzles.

—GORDON HAMILTON

Chapter 13: Problem-Solving

WHEN CHILDREN FACE A TOUGH math problem, their attitude can make all the difference—not so much their "I hate homework!" attitude, but their mathematical worldview. Does your child see math as answer-getting or as problem-solving?

Answer-getting asks "What is the answer?", decides whether it is right, then forgets it and goes on to the next question. Problem-solving cares less about whether an answer is right and more about whether a solution makes sense.

Students who care about problem-solving want to explore the web of interrelated ideas they discovered along the way: How can they recognize this type of problem? Can this one help them figure out others?

What could they do if they had never seen a problem like this one before? How would they reason it out?

Why does the formula work? Where did it come from, and how is it related to basic principles?

What is the easiest or most efficient way to manipulate the numbers? Does this help the problem-solver see more of the patterns and connections within our number system?

Is there another way to approach the problem? How many ways can they think of? Which do they like best, and why?

Task Card Book #1

—37—
Shopping Puzzle

The Horticulture Club has a plant sale: flowering plants are $__ each, and hanging plants are $__ each. [Choose any two numbers.] Sasha has $100 to spend. What might she buy?
  Make up your own shopping puzzle.

—38—
Solve Me Mobiles

Go to solveme.edc.org/mobiles and click "Play." Find a puzzle you like. Copy it in your journal. Explain how you figured it out.
  Make up a mobile puzzle of your own.

—39—
Array Puzzle

Superheroes are lining up in a rectangular array for a parade (or a battle). If they make rows of three, one hero gets left out. If they make rows of five, one hero gets left out. But rows of four work fine. How many superheroes might there be?
  Make up your own lining-up-in-rows puzzle.

—40—
The Mighty Cats

In an ancient Egyptian math puzzle, a rich man's estate contained 7 houses. Every house had 7 cats. Each cat killed 7 mice, which would each have eaten 7 heads of wheat. Every head of wheat, when planted, could produce 7 hekat measures of grain.
  How much grain did the mighty cats save?

Task Card Book #2

—89—
Candy Puzzle

You are working for a chocolate factory. You need to design a box to hold 48 candies. How will you arrange the box? Can you find more than one way?
  What would be the hardest number of candies to package? Why?

—90—
Sci-Fi Puzzle 1

Aliens from the planet Vargo have either 3 or 5 antenna-spikes. [Or choose different numbers.] A Vargon spaceship with a diplomatic crew visited Earth.

When they landed, news reporters counted the total number of spikes, but static interrupted their broadcast. What might the number be? Which numbers are impossible?

—91—
Sci-Fi Puzzle 2

Aliens from the planet Crimbat double their population every year. They send out lots of colony spaceships to keep from overpopulating their home.

  A Crimbatti settlement on Earth contained 6,400 aliens in its 6th year. In which year was the population half that size?
  How many Crimbatti settlers were there in the first year? How long until the aliens outnumber the humans on Earth?

—92—
Same But Different

Go to samebutdifferentmath.com and choose a topic you've studied. Find a puzzle you like and make a sketch of it in your journal.

Compare the two parts. Explain how the images or expressions are the same, and also how they are different from each other.

How many things can you notice?

Task Card Book #3

—141—
Coin Puzzle 1

I'm holding some coins in my hands. Each hand has a different set of coins, but both sets are worth the same amount. How much money might I have?

If I offered you one of the sets of coins, which one would you want?

Make up a new money puzzle of your own.

—142—
Coin Puzzle 2

Pretend you have one of every coin your country makes. What different amounts of money could you make? Are there any amounts you can't make?

Example: a 1-cent and 5-cent coin can make 1, 5, or 6 cents.

What would be the fewest coins you could have that would let you make every amount up to the smallest of your country's paper bills?

—143—
Strange Customers

Mr. Tam sold apples at the Farmer's Market. One morning, each customer bought half of Mr. Tam's apples plus one more, until the seventh customer bought the last two apples. How many apples did Mr. Tam have at the start of the day?

—144—
Fraction Talks

Go to fractiontalks.com and find a puzzle you like. Make a sketch of it in your journal. Identify as many of the fractional parts as you can. Explain how you know the amounts.

Task Card Book #4

—193—
Cookie Bake-Off

If three friends came over for a cookie baking party, and you made 86 cookies in all, how would you share them? Don't forget to give yourself some of the cookies, too!

Make up your own puzzle about cookies.

—194—
Insect Math

Some North American cicadas emerge from the soil to mate every 13 years. Others have a life cycle of 17 years. How often will the two species meet?

Look up some animal facts and make your own math puzzle about them.

—195—
Visual Patterns

Go to visualpatterns.org and find a puzzle you like. Make a sketch of it in your journal. Describe how you see the pattern growing. Use arrows or diagrams as needed.

Pick a number from 50 to 100. Can you tell how many would be in that step of the pattern?

—196—
Percents Puzzle

Tomas runs a computer store. One day he sold two gaming laptops for $1,600 each. He made a 25% profit on the first sale but took a loss of 20% on the second. Overall, did he make a net profit or loss for that day?

Make up a new percents puzzle of your own.

Task Card Book #5

—245—
Goose and Grapes

A goose ate 90 grapes, each day eating 6 more than the day before. How many days did it take? Is there more than one possible answer?

Make up a strange-way-to-count puzzle of your own.

—246—
Which One Doesn't Belong?

Go to wodb.ca and find a puzzle you like. Make a sketch of it in your journal. Write out your answer(s) to the puzzle, using arrows or diagrams as needed.

—247—
Ratio Tables

A ratio table is a problem-solving tool that helps us reason about things that vary in proportion with each other. Think of a "this per that" relationship—eggs per carton, legs per octopus, dollars per gallon.

Make a table showing the relationship, and include some silly entries. How much gas could you buy with a penny? How many eggs in 375 dozen? Or tiny, sharp claws on a million kittens?

Ratio Tables can be used to solve puzzles like this famous brainteaser: A bookworm and a half reads a book and a half in a day and a half. How long does it take for two bookworms to finish two books each?

	book worms	books	days
the riddle	1 ½	1 ½	1 ½
one worm	1	1	1 ½
two worms	2	2	1 ½
two books each	2	4	3

—248—
Championship Math

Ten friends challenge each other to a Rock-Paper-Scissors competition. They face off one-on-one, and each friend plays against everyone else. How many games will they play?

Then they decide that's too confusing, so the loser of each game will be eliminated. Now how many games will it take to determine the R-P-S champion?

Task Card Book #6

—297—
Would You Rather?

Go to wouldyourathermath.com and click your grade range. Find a WYR question you like, and copy it in your journal. Explain your answer: Which would you prefer, and why?

—298—
Domino Counting

How many dominoes are there in a set of double-6 tiles? Each domino has two squares. Each square has 0, 1, 2, 3, 4, 5, or 6 dots. A complete set includes all possible combinations.

Optional challenge: What about a set of double-9 tiles?

—299—
Farmer's Market

Mr. Tam had 480 more pears than apples at his fruit stand. Then he sold half of each type of fruit. Now he has four times as many pears as apples. How many apples did he start with?

Make up a farmer's market puzzle of your own.

—300—
Buckets of Water

The Hobbiton village well has two buckets for drawing water, a large 5-gallon bucket and a smaller one that holds 3 gallons. That makes it easy to get 0, 3, 5, or 8 gallons of water.

What other amounts can you measure out by filling and emptying these buckets? Can you get exactly 4 gallons?

Extra challenge: What if the buckets were different sizes? Can you find another pair of buckets that works to measure any gallon amount?

Mathematics has two faces: it is the rigorous science of Euclid, but it is also something else. Mathematics presented in the Euclidean way appears as a systematic, deductive science; but mathematics in the making appears as an experimental, inductive science. Both aspects are as old as mathematics itself.
—GEORGE PÓLYA

Chapter 14: Experiments

MANY PEOPLE KNOW IT'S IMPORTANT for students to do hands-on experiments in science. But did you know the same is true for mathematics? People learn math by playing with ideas.

A math journal can be like a science lab book. Not the pre-digested, fill-in-the-blank lab books that some curricula provide. But the real lab books scientists write to keep track of their data, and what they've tried so far, and what went wrong, and what finally worked. Children may draw pictures of their investigations, write explanations, or play with equations.

When students find a solution to their prompt question, that's when the fun begins. The point of a math experiment is to change something in the problem and explore how that changes the answer. For older students, the bonus challenge of a math puzzle is to generalize the solution: Can they discover a method that works for any starting conditions?

Besides the prompts in this book, let students pose research topics of their own. What do they wonder? What questions can they ask? Can they expand one of the previous activity prompts into a more general investigation?

Any math topic or prompt offers an overwhelming variety of paths for exploration. Pick your rabbit hole, dive in, and discover the crazy

Wonderland of mathematics.

Be patient with math experiments. Allow plenty of time for students to be scientists, doing their own research. Work for a while and then let the investigation rest. Come back to it later to see if they can discover anything new.

Task Card Book #1

—41—
Linear Crossings

Draw straight lines. Each line must cross all the others. How many lines can you draw? Try the experiment again. Can you get more lines crossing this time? What do you think is the maximum number of straight lines that all cross each other?

—42—
Taxicab Geometry

Draw a rectangular grid to represent your city, with squares big enough to write in. Choose any square for your starting position and label it 0 (zero).

Move from square to square horizontally or vertically, like a taxicab on roads that go east-west or north-south. In taxicab geometry, you never, ever move diagonally. In each square, write the least number of moves (shortest path) to get there from zero.

What patterns do you see? Can you think of any questions to ask?

—43—
Counting Squares

Draw a square array of dots, at least 4 × 4 and up to as large as you wish. Or use dotty graph paper. How many different sizes of squares can you

make by connecting four of those dots? How many squares in all?
Do you want to count tipped or "diamond" squares?
How does the number of squares grow as your array gets bigger? What other questions can you ask?

—44—
Don Steward's Swap

Find a set of numbers A, B, and C that will fit this "swap" equation:

$$A \times (B + C) = A + (B \times C)$$

How many sets can you find?

Task Card Book #2

—93—
A Generous Gift

Pretend your grandparents gave you a penny on the first day of your birthday month, and two pennies on the second day, and kept doubling the pennies until your birthday. How big is that gift?
What if they kept adding pennies for the entire month?

—94—
Quarter the Square

How many different ways can you divide a square into fourths? The four pieces must have the same area, but not necessarily the same shape.
Can you find a division with only one line of symmetry? With no lines of symmetry, but having rotational symmetry? With no symmetry at all?
Sort and classify your designs. What other questions can you ask?

—95—
Exponential Halves

When you tear a piece of paper in half, you get two pieces. If you put one piece on top of the other and tear the stack in half, then you'll have four pieces.

If you were as strong as Superman, you could keep tearing the stack as it gets taller and thicker. How many times would you have to tear the paper to get 64 pieces? What other questions can you ask?

—96—
Exponential Folds

When you fold a piece of paper in half, the result is twice as thick as the sheet by itself. If you fold it again, it will be four times as thick. Another fold makes the entire stack eight times as thick as a single sheet.

If you were able to keep folding the paper over and over, how many folds would it take for the stack to reach as high as the moon?

Task Card Book #3

—145—
Collatz Hailstones

Imagine the whole numbers as tiny ice particles flying in the wind, bouncing up and down inside a storm cloud. Pick any number. If it's even, cut it in half. If it's odd, multiply by 3 and then add 1.

Follow the same rule with your new number, and with each answer after that. If you get down to 1, your original number formed a hailstone that hit the ground.

Try several different starting numbers. What patterns do you see? Can you think of any questions to ask?

—146—
Number Partitions

Choose any positive whole number. Write as many ways as you can think of to express that number by adding positive whole numbers.

Decide whether you want to count what order the *addends* (the numbers you are adding) come in. For example: 5 = 2 + 3 = 3 + 2. Will you count that as two ways to write it, or just one?

Try another starting number. What patterns do you see? Can you think of any questions to ask about number partitions?

—147—
Polygon Diagonals

Draw some polygons (closed shapes with straight sides) and their diagonals. What do you notice about the lines, shapes, and angles formed? What questions can you ask?

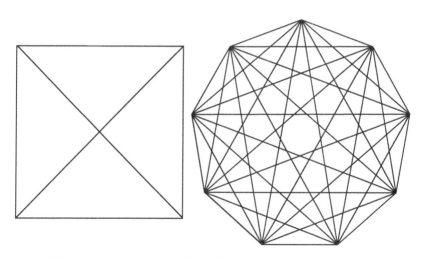

The more corners your polygon has, the more diagonals it will have. How will you know when you have drawn them all?

—148—
Leonhard Euler's Polygon Puzzle

Famous mathematician Euler (pronounced "Oy-ler") wondered how many ways there were to dissect polygons into triangles using their diagonals.

For Euler's puzzle, the diagonal lines can't cross, because crossing might introduce shapes that aren't triangles. And he decided to count all of the mirror images and other rotations that look different from each other, which made the puzzle trickier.

Draw some polygons and divide them with diagonals. How many ways can you find?

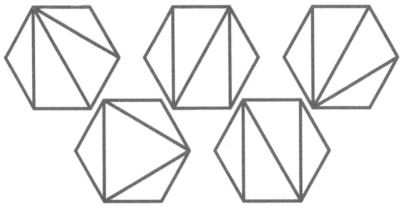

Here are a few example divisions of a hexagon. Notice how none of the diagonal lines cross, and all the interior shapes are triangles.

Task Card Book #4

—197—
Tangrams

Copy the tangram square design (next page) on your page as accurately as possible. What can you say about the angles, sides, and areas of the tangram pieces? Use arrows and diagrams in your explanation, as needed.

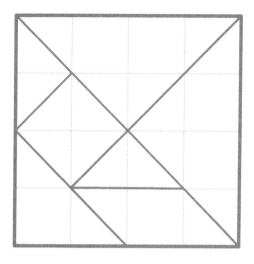

The tangram square.

—198—
Bouncing Billiard Balls

Use graph paper, lined or dotty. Draw any size rectangle for your billiard table. The billiard ball enters at one corner and travels at a 45° angle along the diagonal of each grid square, bouncing off the walls, until it stops at another corner of the board.

Can you predict where the ball will land? Try a variety of different rectangles to look for patterns.

—199—
James Mai's Shape Partitions

Draw a set of 4–8 dots. Copy the exact same set several times on your page. There is only one way to make a single group: just circle all the dots. How many different ways can you make two groups? Show each possibility on one of your dot sets.

Can you divide your dots into three groups? How many ways? What other partitions can you make?

What math do you see in your partition patterns?

What questions can you ask?

Here are a few of the ways to partition six dots arranged in a triangle. Which ones are similar? How are they alike, and how not? What other partitions can you find?

You may enjoy some of James Mai's partition art on the Bridges Mathematical Art Galleries site.†

—200—
Don Steward's Maxagons

Draw several square arrays: 3 × 3 dots, 4 × 4, etc. On each array, connect dots to make a (closed) polygon with the maximum number of sides. How high can you go? Hint: The maxagon for a 3 × 3 dot array has seven sides.

Further investigation: What if you use rectangular arrays such as 3 × 4, 3 × 5, or 3 × any number N? Or 4 × 5, 4 × 6, etc.?

Here are three possible closed polygons on a 3 × 3 dot array. Can you draw one with more sides?

---

† *gallery.bridgesmathart.org/exhibitions/2019-bridges-conference/jlm-phi*
gallery.bridgesmathart.org/exhibitions/2020-joint-mathematics-meetings/jlm-phi

Task Card Book #5

—249—
Digit Sums

Add the digits in a number together. If the sum has more than one digit, repeat until only one digit remains. Does every starting number get down to a single digit?
  What patterns do you see?
  Can you think of any questions to ask?

—250—
Digit Products

Multiply the digits in a number together. If the product has more than one digit, repeat until only one digit remains. Does every starting number get down to a single digit?
  What patterns do you see? Can you think of any questions to ask?

—251—
Dan Finkel's Squarable Numbers

Use square graph paper, lined or dotty. Can you draw a square made of smaller squares? A *squarable number* is a number of squares that can combine to make one larger square.

For example: Try to put two squares together to make a larger square. It's impossible, which means that 2 is not a squarable number. But you can put four same-size squares together to make a bigger square. So 4 is a squarable number.

Another way to think about it: If you can divide a large square into N smaller squares—not necessarily the same size—then N is a squarable number.

How many squarable numbers can you find?

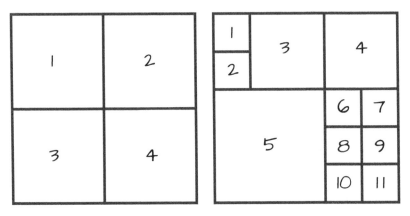

4 and 11 are squarable numbers. Can you find some more?

—252—
Ivan Moscovich's Grasshopper

The grasshopper starts at point 0 on a number line. It makes jumps of 1, 2, 3, …, each jump one unit longer than the previous. Is it possible for the hopper to land on a specific number N after exactly N jumps?

N = 1 works, of course.

Can you find any others?

Task Card Book #6

—301—
Palindromes

A palindromic number reads the same forward or backward, like 121 or 73,537. Choose any number that's not a palindrome. Reverse the digits, and add your new number to your original number. Is the sum a palindrome? If not, repeat the reverse-and-add steps until it is.

Keep track of how many steps it takes to palindromize each number. What patterns do you see?

What questions can you ask?

—302—
Spirolaterals

Use graph paper, lined or dotty. Choose a short list of smallish numbers. For example: 1, 3, 7. Also choose whether to turn right or left (clockwise or counterclockwise).

Draw lines the length of each number, with a right-angle turn at the end of each line. Continue until you get back to your starting point or until your design goes off the page.

With my example list, I would draw a line 1 space, turn, 3 spaces, turn, 7 spaces, turn, 1 space, turn, 3 spaces, and so on until the pattern came back to its beginning.

Explore what happens with different sets of numbers.

—303—
David Butler's Quarter the Cross

Use graph paper, lined or dotty. Draw a cross (plus-sign shape) on graph paper, made of five large squares: one middle square, plus the four squares above and below, left and right. How many grid squares are there in the whole cross?

Color exactly one-fourth of the cross. How do you know you've shaded exactly a quarter? How many different ways can you find to do this?

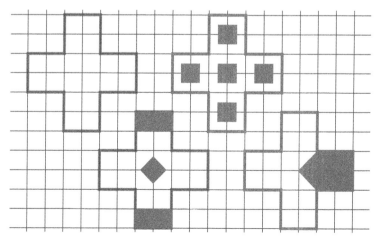

—304—
Pythagorean Tiling

Use graph paper, lined or dotty. Make a repeated, tessellation-style pattern with two different-sized squares, according to this rule: *Every square must touch side-to-side with four squares of the other size, one on each of its sides.*

Each smaller square will be surrounded by four larger ones with no gaps. The larger squares will also touch four others of their own size, forming staggered rows.

Draw additional lines in places that seem interesting on your design. List everything you can see about the lines, angles, symmetries, and patterns formed. What questions can you ask?

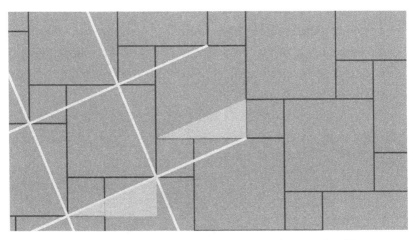

The 9th-century mathematical astronomer Al-Nayrizi "cut" (or *dissected*) the tiling along the light-gray lines to show how the shapes prove the Pythagorean theorem. Can you see that the square on the light-gray triangle's hypotenuse is the sum of the two squares that make the tiling pattern?

Creativity is the heart and soul of mathematics at all levels. To look at mathematics without the creative side of it, is to look at a black-and-white photograph of a Cezanne; outlines may be there, but everything that matters is missing.

—R. C. Buck

Chapter 15: Create Your Own Math

When students create their own math, they forge a personal connection to mathematical concepts and relationships. And it's fun!

Children might make up a math game, write a story or poem, draw a comic, or pose a problem. Create math art, think up a challenging question, or write a puzzle. Since earlier chapters focused on writing and math art, most of these prompts involve creating puzzles or problems.

The "Story Problem Challenge" is one of my favorite math club activities. My students invent their own word problems in any style they like. They don't have to know how to solve the problems they create. We read the stories aloud, and everyone works together to find the solutions.

For puzzles where the child already knows the answers (for example, Two Truths and a Lie), let them trade with a friend. Can they each solve the other's puzzle? Can they stump each other? Or save the child's work and let them come back to it another day, after they've forgotten the answers.

And when students create something they're proud of, let them share it with the world. Visit the Student Math Makers Gallery to learn how your students can submit their own math creations.[†]

---

† *tabletopacademy.net/math-makers*

Task Card Book #1

—45—
Menu Math

Create a menu for an imaginary restaurant. Include main dishes, side dish items, drinks, and desserts. Write a story for your restaurant. What math questions might you ask about your story?

—46—
Age Puzzles

Jewel's father is 3 times as old as Jewel. In 10 years, he will be twice her age. How old is Jewel? Make up some age puzzles of your own.

—47—
Mystery Numbers

Write equations with missing mystery numbers, like

$$3 + \square = 57$$
or
$$(5 \times 5) + 4 = 16 + \square - 3$$

Or use algebra:

$$79 = (n \times 11) + 2$$

Or be silly:

$$\textit{giraffe} \div 4 = 37 - 12$$

Can you solve your mystery number equations? Or trade puzzles with a friend.

—48—
Two Truths and a Lie

Pick a topic you have learned in math. Write two correct statements and one false statement.
    Trade with a friend. Can you find each other's fibs?

—49—
Math Quilt

Draw a rectangular grid of squares on your page, like a quilt made of large, square pieces. For each piece of the quilt, color a different fraction of the square. Or color the same fraction in every square, but make each one look different.
    For example, one square might represent ½, but drawn creatively (not just a line down the middle).

—50—
The Answer Is…

The answer is _____. [Choose any number. Or choose a math vocabulary word.] The question could be…
    Can you think of more than one question? How many possible questions can you find?

—51—
Math Poetry: Limerick

A limerick is a five-line poem, usually comical and sometimes quite rude. Limericks use an AABBA rhyme pattern, with three stressed beats on the A lines and two on the shorter B lines. The rhythm sounds like this:

    da-DUM da-da-DUM da-da-DUM
    da-DUM da-da-DUM da-da-DUM

da-DUM da-da-DUM
da-DUM da-da-DUM
da-DUM da-da-DUM da-da-DUM.

Write a limerick that includes math. For example, here is a limerick about the Collatz conjecture (prompt #145):

A crazy old man, just for fun,
Liked to triple odd numbers, plus one.
"Cut the evens in half,"
He said with a laugh,
Bouncing up, down, up, down, down, down, done.

Optional challenge: Can you figure out which Collatz hailstone number has the bouncing pattern in the last line?

—52—
Reinvent Your Homework 1

Find a page of calculations in your math book, or download a worksheet online. Choose two or three of the questions. Write a story problem to match each calculation.

For example, for the calculation ¾ × 8, you might imagine a recipe that takes ¾ cup of flour. But you are planning a party and need to make eight times that amount. How much flour will you need in all?

Task Card Book #2

—97—
Bus Puzzles

A bus can hold ___ people. It starts out empty (except for the driver). At the first stop, ___ people get on. At the next stop… Write a story for the bus. What math questions might you ask about your story?

—98—
Half Plus Three

Four children get pocket money. Each gets half as much as the next older child, plus $3 more. What questions can you ask?
　　Make up a fraction-plus-a-little-bit puzzle of your own.

—99—
Fictional Math

Think of the characters in your favorite story. How would they use numbers, shapes, or patterns? Would they cook, or go shopping? Might they build something? Would they decorate it with a design? What would they count or measure?
　　Make up some math problems about them.

—100—
Stump an Adult

Write the hardest math problem you can think of. See if your parent or teacher can solve it.

—101—
What Would You Choose?

Would you prefer a stack of quarters equal to your height, or a bag of quarters equal to your weight? Why?
　　Create some make-a-choice questions of your own.

—102—
Ratio Puzzles

The ratio of cats to dogs at the animal shelter is 2.4:1. How many cats and dogs might the shelter have?

What is another possible combination?
Why can't you know the exact answer?
Make up your own ratio puzzles.

—103—
Lines on a Grid

Use dotty graph paper. With a ruler, draw a slanted straight line between two dots on your page.

Can you draw another line parallel to the first?
How do you know it's parallel?
Can you draw a line that's perpendicular? How can you be sure?
Make a design with parallel and perpendicular lines. Color as desired, or fill each section with a pattern.

—104—
Monthly Math

Create a math calendar with a puzzle for each day of the month. Can you make each answer equal the number of that day?

Task Card Book #3

—149—
Number Compositions

Pick a number and see how many ways you can write it. Try to fill your whole page with different expressions for that number.

What kind of crazy math will you create?

For example:

$7 = 5 + 2 = 10 - 3 = \sqrt{49} = (5 + 2) \times (10 - 3) \div (49/7) = \ldots$

—150—
Gadgets Galore

Imagine that you run a hobby shop or gadget store. What do you sell, at what prices? Write a story for your shop. What math questions might you ask about your story?

—151—
Family Math

Write some math problems with questions about your family. Your problems can ask about numbers, shapes, money, time, patterns—or any kind of math you like.

—152—
Today Is…

Write a math problem where the answer is today's date. Can you think of more than one problem? Can you fill a whole page with today-math puzzles?

—153—
Cutting Pizza

Draw a large circle to represent your pizza. Draw straight lines to cut it into slices. Try to make as many pieces as you can. The pieces don't have to be all the same size.

What do you notice? What questions can you ask?

—154—
Permutations

Three students ran a 100-meter sprint. Nobody tied another runner. In what order might they cross the finish line? How many different ways might they finish?

What if there were more runners?

Permutations count how many ways we can arrange things in order—in this case, from first to last. Make up a permutations puzzle of your own.

—155—
Cross-Math Puzzles

Create a mathematical crossword puzzle with clues. The puzzle can use numbers or letters in the squares.

Optional challenge: Make a copy with blank squares plus all the clues. Trade puzzles with a friend.

—156—
Growing Patterns

Draw a pattern that grows according to some rule. Show the first three or four stages of your pattern's life.

Can you describe the growing rule with math?

Examples: visualpatterns.org.

Task Card Book #4

—201—
Animal Emporium

If you managed a store for pets and pet supplies, what would you sell? Would you have any unusual animals who need special care? Would you offer training or other services, and at what rates?

Write a story about your store.

What math questions might you ask?

—202—
Old MacDonald's Farm

Farmer MacDonald is an eccentric old lady. She keeps track of the livestock and poultry on her farm by counting heads and feet.

For example: One day she noticed some sheep and ducks down at the pond with 15 heads and 40 feet. Can you tell how many sheep and how many ducks were there?

Make up your own puzzles about the MacDonald farm.

—203—
Silly Units

Create silly equations using non-standard units of measurement. Share your equations with a friend. Can you guess each other's units?

Examples:

$$2D = 8L + 2T$$
(dogs, legs, tails)

$$2P + 3C = 1F$$
(parents, children, family)

—204—
Reinvent Your Homework 2

Find a page of calculations in your math book, or download a worksheet online. Answer each question math-rebel-style: Write any true statement except what the answer key expects.

Have fun making crazy math.

—205—
Counting Puzzles

If I count by twos, I land on both 100 and 1,000. If I count by threes, I don't hit either 100 or 1,000. Is there any number I can count by to hit only one of them? Make your own counting-by puzzles.

—206—
Magic Math

Pick any number. Add 5 to it. Double that sum. Then subtract 10. Finally, cut that answer in half. What happened?

Can you make up a magic math puzzle? Try to make your puzzle end at the original number, or else go to one particular secret number no matter what the other person chooses to start with.

—207—
Square the Triangle

Use graph paper. How many different triangles can you find that have an area of exactly eight squares?

What other shapes can you find with that area? Can you find a kite? An isosceles trapezoid? An arrowhead? A hexagon? Is any shape impossible?

Make up your own shapes challenge question.

—208—
Invent a Game 1

Make a math board game. Will the players try to conquer territory, or will they race along a path? How will they move? What challenges will you put in their way to make the game more fun?

Try your game with a friend, and tinker with the rules until you're satisfied.

Task Card Book #5

—253—
Business Tycoon

If you could run any kind of business you liked, what would it be? Would you have a shop to sell items? Or would it be a service business where people hire you to do certain tasks? Or both?

Write a story for your business.

What math questions might you ask about your story?

—254—
Cutting a Clock

Using the face of an analog clock as a guide, how many different circle fractions can you draw?

For example: a line from the 12 to the 6 splits the clock in half. What other fractions can you find? How many minutes are there in each fraction slice?

Can you show how to use a clock to add or subtract fractions? Or can you convert the clock-minutes into angle degrees? Or radians? (There are 360° or 2π radians in a full circle.)

What other questions can you ask?

—255—
Math Riddles

Choose a secret number the other players will try to guess. Write a "What Number Am I?" riddle.

For example, "I am odd and prime. I'm a two-digit number less than 30. The sum of my digits is 4. What number am I?"

Give at least three clues for your mystery number. No other number should match all the clues.

For more sample riddles, go to solveme.edc.org/whoami and click "Play."

—256—
Mental Mathstorm

Write a tough calculation in the center of your page. Draw several branches outward. For each branch, write a calculation with easier numbers.

For example, 137 + 359 might have a branch with "130 + 350 + 7 + 9" and another with "130 + 7 + 360 − 1."

Can you fill your whole page with different ways to say the calculation? Which of the variations is the easiest to figure out mentally?

—257—
Invent a Game 2

Can you make a variation of the classic take-away game Nim? What will players take away (or add)? Will they play on a gameboard, or with pieces in a pile? How do you win (or lose) the game?

Try your game with a friend, and tinker with the rules until you're satisfied.

[In Nim, players begin with one or more piles of stones. They remove the stones according to a rule like, "On your turn, take 1–3 stones." Traditionally, the player who takes the last stone loses.]

—258—
Donkey Math

One donkey said to the other, "If I gave you one sack from my load, we'd have the same amount. But if you gave me one sack, I'd be carrying twice as much as you."

What questions can you ask?

Can you make up a donkey math puzzle?

—259—
Four Triangles

Draw a large square. Divide it in half either vertically or horizontally. Then divide those halves along their diagonals to make four right triangles.

How long are the sides of each triangle compared to the original square?

How many other different four-sided shapes (quadrilaterals) can you make with the same set of four triangles?

What other interesting shapes can you make?

Hint: You may want to cut triangles from construction paper to manipulate, then sketch the shapes you find in your journal.

—260—
Vector Algebra

Use graph paper, dotty or lined. "R" means move one space to the right. What do you think "-R" means? "U" means move one space up the page.

What kind of move is "5R - 3U"?

Draw a treasure map, with plenty of obstacles. Use vector algebra to describe a path to the treasure.

Task Card Book #6

—305—
Movie Warehouse

You manage the props department for a movie studio. What is in your inventory? Do you need to buy anything for an upcoming film?

Write a story about your work.

What math questions might you ask about your story?

—306—
Be an Author

Choose a calculation from your math textbook or one that you make up. Write a story about the calculation. Use any genre you like: adventure, fantasy, sports, romance, science fiction, etc.

For example, a subtraction problem might make a story about a nest of dragon eggs where some of the baby dragons hatch and fly away. How many eggs are left?

Can you think of other math questions to ask about your story?

Perhaps my baby dragons came in different colors. So I could ask which color had the most, or how many more green dragons were there than brown ones.

—307—
Animalgebra

Three turtles = 60. One turtle and two fish = 30.

A fish and an octopus = 50.

How much is a turtle, a fish, and an octopus together?

Make up your own animal algebra puzzle. If you like, do it in pictures like a social media meme.

—308—
Invent a Game 3

Create a variation of tic-tac-toe. What size gameboard will you use, and what are the rules for marking squares?

How do you win (or lose) the game?

Is it possible to have a tie?

Try your game with a friend, and tinker with the rules until you're satisfied.

—309—
Make Your Own WODB

Think of three attributes an object might have. For example, you might choose pointy, blue, and striped. Then draw four pictures. One picture should have all three attributes, and the other pictures should each be missing a different one.

Ask a friend, "Which one doesn't belong?"

With a WODB puzzle, each item can be the right answer, as long as you explain why.

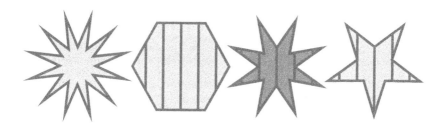

—310—
Threeven Numbers

You've heard of "odd" and "even" numbers. But did you know that numbers can also be "threeven"?

Actually, threeven's not a real word, but when did that ever stop a mathematician? We just make up our own meaning and see where it leads us.

What do you think "threeven" should mean?

Find some threeven numbers using your definition. What patterns do you see?

What questions can you ask about your threeven numbers?

What other mathematical words would you like to make up and define?

—311—
Annie Perkins's Math Art Challenge

Go to arbitrarilyclose.com/home and find a project you like. Play with it. Make your own math art.

What kinds of math can you see in your art?

What questions can you ask?

Take a picture of your art and share it online with the hashtag #MathArtChallenge.

—312—
Lifelong Learning

Create your own list of math research questions. What interesting things would you like to know more about?

You don't have to answer your questions, just wonder about them. Learning to ask good questions is an important part of math.

Section III

Conclusion

Students learn by grappling with mental obstacles and overcoming them. Your students must spend time stuck on problems. The more a teacher steps in to solve a student's problems, the less the student learns.

When a student is working on a hard math problem, they are in a delicate place full of uncertainty, and a lot of the time the ideas they will have are wrong, or at least not exactly right.

One of the best ways to support the creative growth of students is to say yes to their ideas. That doesn't mean confirming the correctness of an idea, but it does mean refraining from pointing out the wrongness.

Instead, encourage students to test out their ideas for themselves.

—Katherine Cook

Thinking is everything in mathematics. Thinking is where joy is to be found. A math class should be an environment where students feel free to share their thinking, and feel no shame about brainstorming.
—FRANCIS SU

Chapter 16: Continue the Adventure

AS OUR WORLD BECOMES EVER more controlled by technology, our children will need ever more math to understand and live in it. Math is not limited to science and tech careers. Business majors study calculus to help them understand growth projections. Artists and musicians need algebra to calculate the return on investment for their promotional expenses. And we all need a good grasp of statistics to make sense of the daily news.

But the math we need is too often not the math our schooling provides. What we need for real life is a true, solid understanding of what I've been calling *real math*, or *math-as-a-nature-walk*. It's a whole different way of thinking.

We need the ability to look at a situation involving numbers, shapes, or patterns and see how the various pieces fit together, how the parts of the situation connect to each other. We need to pay attention to these relationships so we can figure out how changing one bit may affect the whole system. We need to reason in terms of "why?" and "what if?"

I love how math journaling helps students develop true mathematical thinking by making their own discoveries. In many school subjects, children just accept whatever the teacher or textbook says. How do you know about Marco Polo or John Adams or DNA? How do you know that dolphins are mammals or that Jane Austen wrote *Pride and Prejudice?* You know these things only because somebody told you.

But in math, we don't have to just accept what someone tells us. We can figure out things for ourselves.

Like the hiker who encounters a new plant, we can see and touch and wonder about the new ideas we meet. We can create our own responses, and we can communicate our thoughts to others through writing.

Many parents dream of a mathematical magic bullet—a game, app, or book that will help their children learn math and enjoy it. As in life, so also in math: There is no magic solution.

But writing comes close. Writing is not only for language class. It's an important part of every subject, even math, because it's a thinking tool. Writing is very like a mathematician's work, capturing ideas and bringing clarity to what were once vague thoughts.

Before students can put words on paper, they must decide what they think about the subject. As they write, they ask themselves questions, make corrections, and add new insights. They speculate. They engage in critical thinking. They explore a topic or feel their way toward the solution of a problem. And when they read what they have written and discuss it with others, they discover where they skipped steps or have holes in their logic.

When I taught college physics (which is basically applied story problems), the biggest struggle students had was that they didn't know how to use writing as a tool for thinking. They would stare at their homework paper, waiting for the solution to pop into their minds fully formed. If only they had tried some written brainstorming, they'd have found the class much more enjoyable.

Learning math can be hard work, and we need to help our students master every tool available to reason their way through problems. We want them to learn the classic tools of *See, Wonder, Create*. And we want them to know how writing about their ideas can build understanding.

Returning once more to our metaphor of math as a nature walk, sometimes a child will stumble into a boggy area. Progress slows to a muddy slog. And when we see our children struggling like that, we want to build a bridge that passes right over that mucky spot. We want to give them a shortcut, a rule or method that lets them jump from the question straight to the answer without having to get dirty.

For example, we can tell children, "Find a common denominator and then add up the numerators." They can fill out a worksheet. They can get all the right answers without ever coming to grips with what fractions actually mean.

When we try to be helpful, when we offer our students a shortcut bridge—that's fake math. It's not real learning because the students are not reasoning about the problem situation. They are not thinking their own thoughts. We've just given them our thoughts to use as a crutch.

Learning only happens in that boggy mess, when you're in there and getting dirty. That's where you do the mathematical thinking, and you come to understand what's really happening in the problem. That's where you think your own thoughts.

Writing about math requires students to actively confront information. They cannot just passively accept what the teacher or textbook tells them. They must consider the question, organize concepts in their minds, and process it all into knowledge. This helps math stick in a student's mind in a way that merely listening or reading about the topic can never do.

Be a Math Maker

Through the process of getting muddy and messy with math journaling, your children will become Math Makers. They will craft a challenging puzzle, draw a geometric design, tell a story, or invent a new game. Creating math builds deep understanding, as children think about the relationships between numbers, shapes, and patterns. Math becomes personal.

Encourage your children to continue their math-making adventure.

Toys, hobbies, favorite stories—all can be fodder for math creation.

Let the child choose something to think about. Make an "I See" list. How does that item relate to math? What patterns or shapes can you find? Make an "I Wonder" list. How many ways can you turn the things you saw into questions? What else might you ask?

Then it's time to create. Let the student take one of the ideas on their lists and transform it into something new to share with family and friends.

Visit the Student Math Makers Gallery for more information, and to see what other children have made. Then download the "Math Makers Invitation and Submission Form." Send in your child's math creation to share it with viewers all around the world.†

Evergreen Math Prompts

Has your mathematical imagination grown as you explored math journaling? If you want to continue the journey, many of the ideas in this book are designed for repeated play. And here are a few more writing prompts that never grow old:

- ◆ Restate a new concept in your own words.
- ◆ Identify which parts of a lesson you found easy or difficult.
- ◆ Is it true or false that _____? Convince me.
- ◆ Outline the steps to solve a multi-step problem.
- ◆ How might you…
- ◆ Give an example of…
- ◆ How do you feel about…
- ◆ What do you notice…
- ◆ What patterns do you see…
- ◆ What do you wonder…

---

† *tabletopacademy.net/math-makers*

- What is your method for…
- What happens when…
- How do you know…
- What's the hardest (or easiest) thing about…
- Explain how to…
- Explain why…
- Discuss, or reflect on…
- Compare/contrast…
- Define…
- Describe…
- Explore…
- Summarize…
- Analyze…

Be a Math Rebel

Way back when I launched my first math journaling project, I encouraged parents and teachers to look at math education in a new way—to explore math that is creative and even a bit rebellious.

Too often, our children's school math experiences feed the math myths: be fast, get it right, or you're stupid. But a math rebel sees math as a liberal art that focuses on imaginative thinking, not on following school rules.

Math Rebels Believe in Truth

We refuse to accept something just because the teacher or textbook says it. We want to see the connections between math concepts and to understand why things work.

Math Rebels Care About Justice

We resist society's push for speed and conformity. We reject the cultural narrative that math has only One Right Answer.

Math Rebels Celebrate Creative Reasoning

We delight in finding new ways to look at math topics. We want to think deeply about ideas, and we are confident in our ability to figure things out.

As math rebels, we demand the right to make our own math. We know education doesn't belong only in school. Learning happens whenever a curious mind meets an interesting idea. So we yearn for math that meets us each at our own level (justice) and helps us grow in understanding (truth). And we love the creative challenge of inventing our own answers.

Do you want your children to know real math, not just school math? I hope you'll find, as I have, that journaling can be a rewarding part of your children's mathematical adventure.

… and may the Math be with you!

> If math is such an important subject (and it is) why teach it in a way that is dependent on a child's weakest mental ability: memory, rather than her strongest mental ability: imagination?
>
> —Geoff White

Appendixes

Appendix A

21 Favorite Online Resources

Several of the journaling prompts send your student online to choose a math puzzle they like. Those sites are full of great inspiration for mathematical thinking, so I'll list them again here, along with a few other favorites.

Don Steward's Median blog is a veritable treasure house of mathematical prompts for middle school and high school students. So many puzzles, too little time!
donsteward.blogspot.com

Expii Solve poses thought-provoking, interactive math problems (with varying levels of difficulty) that engage students through current events and pop culture.
expii.com/solve

Fractal Kitty shares hands-on math activity ideas, especially for older students. Encourage your students to try one and then write about their experience.
fractalkitty.com/52-weeks-of-math-activities/

Fraction Talks collects images to prompt student curiosity and discussion about whole shapes, parts of the shape, and the numbers we use to represent them.
fractiontalks.com

Graphing Stories demonstrates how a fifteen-second graph can show changing relationships and build understanding.
graphingstories.com

The Julia Robinson Mathematics Festival has collected a fantastic archive of activities "to inspire joy in mathematics through collaboration, exploration, and discovery." Also great for journaling prompts.
jrmf.org/activities

MATH IS FUN offers puzzles, games, definitions, examples, and review lessons for a wide variety of math topics.
mathsisfun.com

MATH MUNCH highlights all sorts of creative ways to play with math: stories, games, videos, puzzles, art, interviews, and lots more. Browse and enjoy.
mathmunch.org

MATH PICKLE offers an amazing collection of original puzzles, games, and student-friendly unsolved math problems. Yes, "unsolved," as in mathematicians don't know the answers, yet these problems are suitable for child's play.
mathpickle.com

MATH TALKS (including PATTERN TALKS) inspires students to apply whatever they've learned to make sense of math questions. A nice collection of number- and shape-based discussion prompts with sample student answers.
mathtalks.net

THE #MATHARTCHALLENGE demonstrates one hundred artistic ways to engage students with rich mathematical ideas. These activities span a wide range of topics, from arithmetic to polygons to graph theory and beyond.
arbitrarilyclose.com/home

MATHIGON is building a fun collection of math activities you can play online, plus interactive lessons and resources for further study. Not to mention a beautiful timeline of math history.
mathigon.org

NRICHMATH is a delightful site full of playful math ideas and investigations to develop mathematical thinking and problem-solving skills in students of all ages.
nrich.maths.org

OPEN MIDDLE puzzles offer multiple ways to approach and ultimately solve the problem. Great for learners of any skill level.
openmiddle.com

THE PARALLEL MATH PROJECT posts a set of weekly math challenges: mystery and history, activities and oddities, puzzles and problems for middle and high school students.
parallel.org.uk

SAME BUT DIFFERENT MATH features visual prompts to launch mathemat-

ical discussions or short essays about the meaning of equivalence in topics from early counting to algebra.
samebutdifferentmath.com

S<small>LOW</small>-R<small>EVEAL</small> G<small>RAPHS</small> and W<small>HAT'S</small> G<small>OING</small> O<small>N IN</small> T<small>HIS</small> G<small>RAPH</small>? encourage sense making and critical thinking about how data tell a story and how graphs make that story visible.
slowrevealgraphs.com
nytimes.com/column/whats-going-on-in-this-graph

S<small>OLVE</small>M<small>E</small> contains logic puzzles that help build algebraic thinking. Play with the online puzzles or build your own creations.
solveme.edc.org

V<small>ISUAL</small> P<small>ATTERNS</small> helps students understand functions and pave the way for algebra. The puzzles are a wonderful way to prompt mathematical discussions as students share and compare ideas.
visualpatterns.org

W<small>HICH</small> O<small>NE</small> D<small>OESN'T</small> B<small>ELONG</small>? puzzles ask students to notice details and compare attributes. Every answer is correct, as long as you explain why.
wodb.ca

W<small>OULD</small> Y<small>OU</small> R<small>ATHER</small>? questions encourage estimation and reasoning about math in real-life situations and help students use math to justify their ideas.
wouldyourathermath.com

APPENDIX B

Quote and Reference Links

ALL THE WEBSITE LINKS IN this book were checked before publication, but the Internet is volatile. If a website disappears, you can run a browser search for the author's name or article title. Or try entering the web address at the Internet Archive Wayback Machine.
archive.org/web/web.php

BERLINGHOFF, WILLIAM AND FERNANDO GOUVÊA. "It is all too common for students to experience school mathematics..." from *Math Through the Ages: A Gentle History for Teachers and Others,* second edition, Dover, 2019 (AMS/MAA Press, 2015).

BOGART, JULIE. "Natural Stages of Growth in Writing," Brave Writer website. Bogart is the author of *The Writer's Jungle.*
bravewriter.com/getting-started/stages-of-growth-in-writing

BUCK, R C. "Creativity is the heart and soul of mathematics..." quoted by John A Brown and John R. Mayor in "Teaching Machines and Mathematics Programs," *The American Mathematical Monthly,* v 69 n 6, June–July 1962.
jstor.org/stable/2311205
https://www.jstor.org/stable/2311205

BURNS, MARILYN. "Writing in math class isn't meant to produce a product suitable for publication..." from "Writing in Math." *Educational Leadership,* October 2004. Burns's blogs, podcasts, videos, articles, and more are available at marilynburnsmath.com. She tweets at @mburnsmath.
marilynburnsmath.com/articles/WritinginMath.pdf

BUTLER, DAVID. "Digit Disguises," Making Your Own Sense blog, September 21, 2019. Butler is a lecturer in the Maths Learning Centre at the University of Adelaide, Australia.
blogs.adelaide.edu.au/maths-learning/2019/09/21/digit-disguises

—. "A Day of Maths 2: Quarter the Cross," Making Your Own Sense blog,

July 5, 2016.
blogs.adelaide.edu.au/maths-learning/2016/07/05/a-day-of-maths-2-quarter-the-cross

———. "Quarter the Cross," Making Your Own Sense blog, April 12, 2016.
blogs.adelaide.edu.au/maths-learning/2016/04/12/quarter-the-cross

COOK, KATHERINE. "Students learn by grappling with mental obstacles…" from "5 principles of extraordinary math teaching," Math for Love blog, June 13, 2015. Cook is the Creative Director at Math for Love, a great resource for creative math ideas and activities.
mathforlove.com/2015/06/5-principles-of-extraordinary-math-teaching

DANIELSON, CHRISTOPHER. "The question, 'What did you learn?' implies the process has ended…" from "What Did You Learn?", Overthinking my teaching blog, August 23, 2013. Danielson is the author of *Which One Doesn't Belong?* and *How Many?* and writes the Talking Math with Your Kids blog.
christopherdanielson.wordpress.com/2013/08/23/what-did-you-learn
talkingmathwithkids.com

FINKEL, DAN. "Mathematics is not about following rules…" from "5 Ways to Share Math with Kids," TEDxRainier, November 2015. Finkel is a founder of Math for Love.
ted.com/talks/dan_finkel_5_ways_to_share_math_with_kids
mathforlove.com

———. "Broken Calculators," Math for Love site.
mathforlove.com/lesson/broken-calculators

———. "Squarable Numbers," Math for Love site.
mathforlove.com/lesson/squarable-numbers

———. "Subtracting Reverses," Math for Love site.
mathforlove.com/lesson/subtracting-reverses

FRANCO, BETSY. *Mathematickles,* Margaret K. McElderry Books, 2006.

FULTON, BRAD, AND BILL LOMBARD. "Games and Puzzles that Reach the Kids and Teach the Standards," notes from the California Math Council conference at Asilomar, December 2009.
tinyurl.com/games-and-puzzles-handout (PDF)

GARDNER, MARTIN. "Sim, Chomp and Race Track: New Games for the Intellect (and Not for Lady Luck)," Mathematical Games column in *Scientific American* magazine, January 1973.

GASKINS, DENISE. *Let's Play Math: How Families Can Learn Math Together, and Enjoy It,* Tabletop Academy Press, 2016.

—. "The Best Math Game Ever," Let's Play Math blog, August 25, 2020.
denisegaskins.com/2020/08/25/the-best-math-game-ever

—. "Dot Grid Doodling," Let's Play Math blog, March 23, 2017.
denisegaskins.com/2017/03/23/dot-Grid-doodling

—. "Math & Education Quotations," Let's Play Math website.
denisegaskins.com/best-of-the-blog/quotations

—. "Math Journals and Creative Reasoning," Let's Play Math blog, January 28, 2021.
denisegaskins.com/2021/01/28/math-journals-and-creative-reasoning

—. "Math Journals for Elementary and Middle School," Let's Play Math blog, August 24, 2018.
denisegaskins.com/2018/08/24/math-journals-for-elementary-and-middle-school

—. "Year Game," Let's Play Math website.
denisegaskins.com/tag/year-game

GAUSS, KARL FRIEDRICH. "It is not knowledge, but the act of learning..." from a letter to János Bolyai, 1808. Quoted at the MacTutor History of Mathematics Archive.
mathshistory.st-andrews.ac.uk/Biographies/Gauss/quotations

GOLDEN, JOHN. "Games help set the culture I want to develop..." from "Yes, Playing Around," *Mathematics Teaching in the Middle School* website, August 29, 2014. Quoted at DeniseGaskins.com. Golden helps train future math teachers as an associate professor at Grand Valley State University, and trains the rest of us through the posts on his blog.
web.archive.org/web/20210125130752/https://www.nctm.org/Publications/
    Mathematics-Teaching-in-Middle-School/Blog/Yes,-Playing-Around
denisegaskins.com/2020/11/14/math-as-a-verb/

—. "There's something about this assignment..." and fraction square images, from Twitter feed for @mathhombre.
twitter.com/mathhombre/status/1343255683316199424
twitter.com/mathhombre/status/1337603004619501569

—. "Knot Fun," Math Hombre blog, March 17, 2011.
mathhombre.blogspot.com/2011/03/knot-fun.html

—. "Some Sum to One," Math Hombre blog, February 27, 2012.
mathhombre.blogspot.com/2012/02/some-sum-to-one.html

GOUVÊA, FERNANDO AND WILLIAM BERLINGHOFF. "It is all too common for students to experience school mathematics…" from *Math Through the Ages: A Gentle History for Teachers and Others*, second edition, Dover, 2019 (AMS/MAA Press, 2015).

HAMILTON, GORDON. "The experience of mathematics should be profound and beautiful…" from "About Math Pickle," Math Pickle website. Hamilton posts games and activity ideas for all ages.
mathpickle.com/about-mathpickle

JORIS, WALTER. Sequencium game from Ben Orlin's "Six Strategic Pen-and-Paper Games (from a Strange and Bottomless Mind)," Math with Bad Drawings blog, April 22, 2020. Joris is an artist and the author of *100 Strategic Games for Pen and Paper*.
*mathwithbaddrawings.com/2020/04/22/
  six-strategic-games-from-a-strange-and-bottomless-mind*
waljoris.blogspot.com/p/biography.html

KIEFER, KATE, AND MIKE PALMQUIST, NICK CARBONE, MICHELLE COX, DAN MELZER. "An Introduction to Writing Across the Curriculum," The WAC Clearinghouse, 2000–2021.
wac.colostate.edu/resources/wac/intro

KRIEG, PAULA BEARDELL. "How to Make an Origami Pamphlet," Playful Bookbinding and Paper Works blog, November 30, 2009. Artist Paula Beardell Krieg makes works-on-paper, teaches book arts, and shares her thinking about math.
bookzoompa.wordpress.com/2009/11/30/how-to-make-an-origami-pamphlet

LAIB, JENNA. "The Simple-but-High-Leverage Game Collection: Making Games Routine," Embrace the Challenge blog, May 28, 2019. Laib works as a K–8 math specialist near Boston.
jennalaib.wordpress.com/2019/05/28/the-simple-but-high-leverage-game-collection

LOCKHART, PAUL. "Mathematical reality is an infinite jungle full of enchanting mysteries…" from *Measurement*, Harvard University Press, 2012. Lockhart is also the author of *A Mathematician's Lament: How School Cheats Us Out of Our Most Fascinating and Imaginative Art Form*.

—. "The mathematical question is…" from the video trailer for his book

Measurement, Harvard University Press, 2012.
youtu.be/V1gT2f3Fe44

LOMBARD, BILL, AND BRAD FULTON. "Games and Puzzles that Reach the Kids and Teach the Standards," notes from the California Math Council conference at Asilomar, December 2009.
tinyurl.com/games-and-puzzles-handout (PDF)

MAI, JAMES. "Shape-Partitions: New Elements, New Artworks," Bridges 2019 Conference Proceedings. Mai is a professor at the Illinois State University School of Art. You may also enjoy his artwork from the Bridges Art Galleries.
archive.bridgesmathart.org/2019/bridges2019-171.pdf
gallery.bridgesmathart.org/exhibitions/2019-bridges-conference/jlm-phi
gallery.bridgesmathart.org/exhibitions/2020-joint-mathematics-meetings/jlm-phi

MERCAT, CHRISTIAN. "Celtic Knotwork: The ultimate tutorial," Entrelacs.net site, March 27, 2019.
entrelacs.net/-celtic-knotwork-the-ultimate

MEYER, DAN. "Tiny Math Games," dy/dan blog, April 16, 2013.
blog.mrmeyer.com/2013/tiny-math-games

MOSCOVICH, IVAN. *The Puzzle Universe: A History of Mathematics in 315 Puzzles,* Firefly Books, Reprint edition, 2019.

NRICH TEAM. "Attractive Tablecloths," Nrich Enriching Mathematics website.
nrich.maths.org/900

—. "Maxagon," Nrich Enriching Mathematics website.
nrich.maths.org/11236

OLLERENSHAW, KATHLEEN. "Mathematics is a way of thinking…" from her autobiography *To Talk of Many Things,* Manchester University Press, 2004. Quoted by Joe Shervin in "The remarkable life of Dame Kathleen Ollerenshaw," The Hub, August 22, 2019.
mub.eps.manchester.ac.uk/science-engineering/2019/08/22/
the-remarkable-life-of-dame-kathleen-ollerenshaw

ORLIN, BEN. "If this feels hard, that doesn't mean you're a failure…" from "Learning to rock-climb is changing how I'll teach math," Math with Bad Drawings blog, October 30, 2013. Orlin is the author of *Math with Bad Drawings* and *Change is the Only Constant.*
mathwithbaddrawings.com/2013/10/30/
im-learning-to-rock-climb-its-changing-how-ill-teach-math

———. "Six Strategic Pen-and-Paper Games (from a Strange and Bottomless Mind)," Math with Bad Drawings blog, April 22, 2020.
mathwithbaddrawings.com/2020/04/22/
  six-strategic-games-from-a-strange-and-bottomless-mind

PARKER, RUTH. "I used to think my job was to teach students…" from "The Having of Mathematical Ideas: Learning to Listen to Students," presentation at NCSM Annual Conference, 2013. Quoted by Joe Schwartz in "Then and Now." Exit 10a blog, January 5, 2017.
exit10a.blogspot.com/2017/01/then-and-now.html

PERKINS, ANNIE. Arbitrarily Close blog, home of the #MathArtChallenge, the Mathematician Project, and "Links to Resources on Not Just White Dude Mathematicians." Perkins teaches middle and high school math.
arbitrarilyclose.com/home
arbitrarilyclose.com/mathematician-project
arbitrarilyclose.com/links-to-resources-on-not-just-white-dude-mathematicians

PÓLYA, GEORGE. "Mathematics has two faces…" from the preface of *How to Solve It*, Princeton University Press, 1945.
mathshistory.st-andrews.ac.uk/Extras/Polya_How_to_solve_it

POST, SONYA. "Substitution Game—Forget the Worksheets," Arithmophobia No More blog, 2016. Former "math witch" and author of *Hands-On Learning with Gattegno*, Post equips homeschooling parents to understand and teach math.
arithmophobianomore.com/substitution-game-forget-worksheets

REINHART, STEVE. "When I was in front of the class…" from "Never Say Anything a Kid Can Say," *Mathematics Teaching in Middle School*, v5 n8, April 2000.
georgiastandards.org/resources/Online%20High%20School%20Math%20Training%20
  Materials/Math-I-Session-5-Never-Say-Anything-a-Kid-Can-Say-Article.pdf

REISCH, GREGOR. "Arithmetica," woodcut, 1503, public domain. Lady Arithmetica (the muse of Arithmetic) watches over a contest between one man (perhaps Boethius) using the new-to-Europe Hindu-Arabic numbers in a written algorithm and another (perhaps Pythagoras) using a traditional counting-board abacus.

SALLAY, IVA. Find the Factors blog, the home of one of my all-time favorite multiplication puzzles.
findthefactors.com

SANDERS, SAVANNAH. "Math is not just adding, subtracting, multiplying, and dividing..." from "Never Give Up," Black Women Rock Math blog, October 16, 2020. Sanders is a nine-year-old (when she wrote her article) mathematician, gymnast, dancer, and girl scout.
blackwomenrockmath.com/blog/f/never-give-up

SHAH, MANAN. "Alphanumeric Puzzle 8—Two Wrongs Don't Make a Right, Five Do," Math Misery blog, January 19, 2017. Shah posts about math, education, and life on twitter as @shahlock.
mathmisery.com/wp/2017/01/19/alphanumeric-puzzle-8

SMITH, DAVID EUGENE. "What would mathematics have amounted to without the imagination of its devotees..." quoted by Rosemary Schmalz in *Out of the Mouths of Mathematicians: A Quotation Book for Philomaths,* American Mathematical Society, 2020.

STANLEY, RICHARD, AND ERIC W WEISSTEIN. "Catalan Number." From MathWorld: A Wolfram Web Resource.
mathworld.wolfram.com/CatalanNumber.html

STEWARD, DON. Median blog. A wonderful resource for math puzzles, especially for middle school and older students.
donsteward.blogspot.com

—. "Algebra snakes and branches," Median blog, January 5, 2018.
donsteward.blogspot.com/2018/01/algebra-snakes-and-branches.html

—. "Area for a fixed perimeter," Median blog, December 1, 2015.
donsteward.blogspot.com/2015/12/area-for-fixed-perimeter.html

—. "Area is 8 squares," Median blog, November 15, 2015.
donsteward.blogspot.com/2015/11/area-8.html

—. "Comparing two data sets," Median blog, January 21, 2012.
donsteward.blogspot.com/2012/01/comparing-two-data-sets.html

—. "Four triangles," Median blog, March 29, 2012.
donsteward.blogspot.com/2012/03/four-triangles.html

—. "Maxagon," Median blog, April 5, 2012.
donsteward.blogspot.com/2012/04/maxagon.html

—. "Number pyramids," Median blog, May 20, 2015.
donsteward.blogspot.com/2015/05/pyramids.html

—. "Quadrilaterals on a 3 by 3 dotty grid," Median blog, August 17, 2017.
donsteward.blogspot.com/2017/08/quadrilaterals-on-3-by-3-grid.html

—. "Swap times with sum," Median blog, April 27, 2010.
donsteward.blogspot.com/2010/04/swap-times-with-sum.html

SRINIVASAN, BHAMA. "Truth and beauty are enough…" quoted in "Bhama Srinivasan: The Invasion of Geometry into Finite Group Theory," Association for Women in Mathematics site, 2005.
web.archive.org/web/20160303173824/https://awm-math.org/noetherbrochure/Srinivasan90.html

SU, FRANCIS. "Thinking is everything in mathematics…" from "Teach math like you'd teach writing," Francis Su's blog, August 24, 2020. Su is the author of *Mathematics for Human Flourishing*.
francissu.com/post/teach-math-like-youd-teach-writing

TANTON, JAMES. "This is the wonderful thing about just thinking and playing…" from "Math and Cats," National Math Foundation newsletter, September 14, 2020. Tanton is an author, former teacher, and one of the founders of the Global Math Project.
globalmathproject.org/wp-content/uploads/2020/01/Issue-1_-NMF-Weekly-Newsletter_-Math-and-Cats.pdf

THOMPSON, CHRISTY HERMANN AND KASSIA OMOHUNDRO WEDEKIND. "Math is not neutral, nor objective…" from "Students as Reasoners in the Hands-Down Conversation," Hands Down, Speak Out blog, August 12, 2020.
handsdownspeakout.wordpress.com/2020/08/12/students-as-reasoners-in-the-hands-down-conversation

VANDERVELDE, SAM. "Game of Criss-Cross," Math Teachers' Circle Network website, January 10, 2020.
mathteacherscircle.org/session/game-criss-cross

WEDEKIND, KASSIA OMOHUNDRO AND CHRISTY HERMANN THOMPSON. "Math is not neutral, nor objective…" from "Students as Reasoners in the Hands-Down Conversation," Hands Down, Speak Out blog, August 12, 2020.
handsdownspeakout.wordpress.com/2020/08/12/students-as-reasoners-in-the-hands-down-conversation

WEISSTEIN, ERIC W, AND RICHARD STANLEY. "Catalan Number," from

MathWorld: A Wolfram Web Resource.
mathworld.wolfram.com/CatalanNumber.html

WHITE, GEOFF W. "If math is such an important subject..." from "The Grade 10 Math Crunch, or Hitting the Wall at Grade 10," Teaching Math with Manipulatives website.
geoffwhite.ws/hittingthewall.html

WIKIPEDIA CONTRIBUTORS. "Chomp," Wikipedia Internet Encyclopedia.
wikipedia.org/wiki/Chomp

—. "Collatz conjecture," Wikipedia Internet Encyclopedia.
wikipedia.org/wiki/Collatz_conjecture

—. "Gomoku," Wikipedia Internet Encyclopedia.
wikipedia.org/wiki/Gomoku

—. "Lune of Hippocrates," Wikipedia Internet Encyclopedia.
wikipedia.org/wiki/Lune_of_Hippocrates

—. "Nim," Wikipedia Internet Encyclopedia.
wikipedia.org/wiki/Nim

—. "Pig (dice game)," Wikipedia Internet Encyclopedia.
wikipedia.org/wiki/Pig_%28dice_game%29

—. "Spirolateral," Wikipedia Internet Encyclopedia.
wikipedia.org/wiki/Spirolateral

—. "Wythoff's game," Wikipedia Internet Encyclopedia.
wikipedia.org/wiki/Wythoff%27s_game

WILLIAMS, DAVID R. "Earth Fact Sheet," NASA Goddard Space Flight Center, November 25, 2020.
nssdc.gsfc.nasa.gov/planetary/factsheet/earthfact.html

WILLIAMS, FARRAR. *Numberless Math Problems: A Modern Update of S.Y. Gillians Classic Problems Without Figures,* self-published, 2018.

—. "Math With No Numbers," I Capture the Rowhouse blog, September 4, 2018.
farrarwilliams.wordpress.com/2018/09/04/math-with-no-numbers

ZINSSER, WILLIAM. "Writing organizes and clarifies our thoughts..." from *Writing to Learn,* HarperCollins Publishers, 1988.

APPENDIX C

Acknowledgments and Credits

A HUGE "THANK YOU!" to my beta readers: Adrienne, Aleena, Amber, Angel, Angela, Fatima, Janine, Ly-ann, Marie-Pier, Miranda, Natalie, Ri, and Susan. Your comments improved the book tremendously.

And to my wonderful editor, Robin Netherton: I'm always amazed at how many mistakes you catch after I think my work is done. Thank you! (Any remaining errors are due to my continued tinkering after the file left her capable hands.)

Cover photo by Sharomka via depositphotos.com.
depositphotos.com/390520992/stock-photo-educate-home-child-girl-make.html

"Arithmetica" by Gregor Reisch, 1503, public domain.
*maa.org/press/periodicals/convergence/
mathematical-treasures-margarita-philosophica-of-gregor-reisch
commons.wikimedia.org/wiki/File:Gregor_Reisch_-_Margarita_Philosophica_-_
Arithmetica.jpg*

"Find the Factors" puzzles by Iva Sallay. Used by permission.
findthefactors.com

"Roman Geometric Mosaic" from the "Palazzo Massimo alle Terme" National Roman Museum of Rome, Italy. Photo by Mbellacini via Wikimedia Commons, CC BY-SA 4.0.
commons.wikimedia.org/wiki/File:Roman_geometric_mosaic.jpg

"How to Make an Origami Pamphlet" illustration by Paula Beardell Krieg, Playful Bookbinding and Paper Works blog, November 30, 2009. Used by permission.
bookzoompa.wordpress.com/2009/11/30/how-to-make-an-origami-pamphlet

Fraction square images, from John Golden's Twitter feed for @mathhombre. Used by permission.
twitter.com/mathhombre/status/1337603004619501569

"Two cats" sketch by Fennywiryani. "Girl with notebook" sketches by OlgaTropinina. "Geometric low-poly bear" illustration by Vea-blackfox. Assorted doodle sketches by Bioraven, HitToon, MisterElements, UncleLeo, Olga.angelloz, Romanchik, Sashatigar, Seamartini, Toponium, and Undrey. All via depositphotos.com.

Quotation background photo courtesy of Loren Joseph on Unsplash.
unsplash.com/photos/By9GOo49sPo

Author photo by Christina Vernon.
melliru.com

Index

2-D Nim, 41

A
acrostic, 94
Adams, John, 166
addition
    game, 39, 41, 42, 44, 46, 54
    math facts, 57
    pattern, 53, 60, 61, 141
    puzzle, 54, 62
age puzzles, 150
algebra
    averages, 123
    code game, 40
    equations, 150, 157
    geometric relationships, 64
    number patterns, 51
    puzzle, 139, 162
    skip-counting, 48
    snakes, 58
    vectors, 161
Alhazen, lunes of, 74
aliens, 103, 130
Al-Nayrizi, 148
alphametrics, 60
angles
    in a circle, 28
    in a hexagon, 68
    on a clock, 122, 159
animal algebra, 162
answer-getting, 128
answers
    as performance, 14
    how to check, 22, 107

answers (*continued*)
    making sense, 3
    math-rebel style, 32, 157
    more than one path, 22
    no answer key, 22, 27
    shortcut, 168
    side-effect of reasoning, 27
    speed, 8
    the answer is, 151
apples, 119, 132, 135
Archimedes, 8, 64
area
    fractions, 139, 147, 151
    in a circle, 73
    maximum, 69
    multiplication as, 41
    puzzle, 65, 67, 70, 72, 158
    squares and roots, 65, 66
Arithmetica, 6, 7, 182
arrays, 33, 129
art, 7
    how to use, 11, 75
    Math Art Challenge, 164, 175
    asking new questions, 37
Austen, Jane, 167
automathography, 90, 119
average, 120, 122, 123, 124, 125, 127
Avoid Three, 49

B
Banneker, Benjamin, 111
beauty, 75, 109, 128
benchmark measurements, 125
Berlinghoff, William, 110, 177

billiards, 143
binary numbers, 113
biography
　automathography, 90, 119
　research, 111, 112, 113, 115, 116, 117
Blockout, 41
Bogart, Julie, 24, 177
Bowling, 39
brainstorming, 20, 25, 54, 96
brainteaser, 104, 134
Brave Writer, 25, 177
Bridges Mathematical Art Galleries, 144
broken calculator, 55, 56
Buck, R.C., 149, 177
Burns, Marilyn, 30, 177
Butler, David, 147, 177

C
calculator, 55, 56, 107
calendar, 154, 155
candy, 43, 104, 123, 130
cardioid, 84
cards, 38, 51, 54
career math, 111, 115, 166
cats, 1, 129, 153
　escape puzzle, 2
　princess puzzle, 3
Celtic knot designs, 117
Cezanne, 149
chessboard, 85, 88
Chomp, 43
chords, 35
cicadas, 132
circles
　analyzing a prompt, 27
　diameter, 73
　lunes, 73, 74
　tangent, 71
　using a compass, 76
　vocabulary, 34
classic puzzles
　age conundrums, 150
　billiards, 143
　buckets of water, 136
　clock puzzle, 122

classic puzzles (*continued*)
　Collatz conjecture, 140, 152
　crossing lines, 138
　cutting a plane, 155
　donkey math, 160
　doubling, 88, 139
　Earth's belt, 114
　Egyptian cats, 129
　four 4s, 63
　half plus one more, 132
　half that size, 130
　heads and feet, 157
　magic math, 158
　number partitions, 141
　one left out, 129
　palindrome numbers, 146
　polygon dissections, 142
　Pythagorean dissection, 148
　rate conundrum, 134
　squaring the circle, 73
　stacks of paper, 140
　surprise, 107
　tangrams, 142
cliché, 60
clock fractions, 159
clock puzzle, 122
codes, 40
coin puzzles, 102, 131
Collatz conjecture, 140, 152, 185
comic strip, 79
Connect 4, 42
connecting dots
　game, 40, 43
　math art, 78
　maxagons, 144
　maze, 82
　octagons, 78
　puzzle, 71
　quadrilaterals, 69, 72
　squares, 138
　tessellations, 83
　with a ruler, 84, 154
Cook, Katherine, 165, 178
cookbook, 16
cookie bake-off, 132

corrections, 29
counting squares, 138, 145
create a font, 76
creativity, 20, 149, 171
Criss-Cross, 40
criticism, 28, 165
crossing lines, 138
cross-math puzzles, 156
cryptarithmetic, 60
Cuisenaire rods, 79
curiosity, 121
   and learning, 20, 37, 98, 171
   posing questions, 3, 93, 137, 164, 169
curiosity, 172
cutting pizza, 155

D

Danielson, Christopher, 37, 178
data, 118
   averaging, 124, 127
   graphing, 121, 122, 123, 124, 126, 127
   in the news, 116
   life expectancy, 113
   supporting, 117
diagonals, 141, 142
dialogue journal, 30
diameter, 27, 34, 73, 74, 93
dice, 38, 39, 41, 44, 46, 54, 125
dominoes, 135
donkey math puzzle, 160
doodling, 75, 79, 84
Double Digit, 44
doubling, 54, 88, 130, 139
drafting, 76, 84

E

Earth's belt, 114
editing, 28
Egyptian math, 112, 129
Einstein, Albert, 5
equator, 114
estimation, 115, 126
Euclid, 137
Euler, Leonhard, 142
experiments, 2, 38, 137
Expii Solve, 104, 174

exponents, 55, 59, 140

F

factorials, 124
fake math, 168
farmer's market, 132, 135
fear of math, 2, 7
feelings, 7, 12, 15, 98
Fermat's Last Theorem, 22
Fib poem, 92
Fibonacci, 60, 92
Find the Factors, 57, 182
Finkel, Dan, 5, 56, 145, 178
following rules, 9, 19, 33, 51
font, create a, 76
four 4s, 63
Fraction Talks, 132, 174
fractions
   common denominator, 168
   of a square, 121, 139, 151
   on a clock, 159
   puzzles, 132, 147, 153
   simplify, 100
   wall, 119
Franco, Betsy, 90, 178
Frayer model, 89
freewriting, 96

G

games
   analyzing strategy, 38, 105
   changing the rules, 38, 158, 160, 162
   freewrite, 97
   how to use, 10, 38
   Math You Can Play, 17
   misère, 38
games, by name
   2-D Nim, 41
   Avoid Three, 49
   Blockout, 41
   Bowling, 39
   Chomp, 43
   Connect 4, 42
   Criss-Cross, 40
   Double Digit, 44
   Gomoku, 45

games, by name (*continued*)
　Greedy Pig, 46
　Make a Square, 46
　Nim, 39, 45
　One Hundred Up, 42
　Pig, 41
　Place Value Nim, 46
　Row Call, 49
　Secret Number Codes, 40
　Sequencium, 49
　Sim, 43
　Skip-Counting Game, 48
　Substitution Game, 42
　Tsyanshidzi, 45
　Wythoff's Nim, 47
　Year Game, 56
Gardner, Martin, 112, 178
Gauss, Karl Friedrich, 8, 110, 179
Geometria, 7
geometry
　analyzing a prompt, 27
　how to use, 64
　vocabulary, 33
Gillian, S.Y., 104
Golden, John, 38, 121, 179
Gomoku, 45
goose and grapes, 133
Gouvêa, Fernando, 110, 180
government, 113, 127
grammar, 30
graph paper, 18
graphing, 118
　data, 121, 122, 123, 124, 126, 127
　in the news, 116
grasshopper numbers, 146
Greedy Pig, 46
grids, 33
growth mindset, 19, 88

H
haiku, 89
hailstone numbers, 140, 152
Hamilton, Gordon, 128, 180
hexagons, 34, 68, 142
hierarchy of editing, 28

hieroglyphic numerals, 112
Hippocrates, 73, 185
Hippocrates, lunes of, 73
history, 110, 114, 115, 116
hobbits, 32, 136
homework
　answer-getting, 128
　math-rebel-style, 32
　prompt, 152, 157
　struggles, 7, 167
hundred face, 79

I-J-K
imagination, 20, 94, 96, 172
Incompetech, 18
insect puzzle, 132
interior decorating, 122
Joris, Walter, 49, 180
Kickstarter, 1, 31
Krieg, Paula Beardell, 105, 180

L
Lady Arithmetica, 6, 7
Lady Geometria, 7
Laib, Jenna, 48, 180
learning tools, 19, 167
liberal arts, 6, 7, 16, 170
life expectancy, 113
lifelong learning, 4, 30, 37, 164, 171
limerick, 151
listening to children, 15, 21
living math, 17
Lockhart, Paul, 8, 102, 180
Lovelace, Ada, 112
low poly art, 81
lunes, 73, 74

M
MacTutor Biographies, 111, 112, 113, 115, 116, 117
magic math puzzle, 158
Mai, James, 143, 181
Make a Square, 46
makers, 4, 20, 93, 149, 168
mandala, 85
maps, 100

Math Art Challenge, 164, 175, 182
math concepts
   addition, 39, 41, 42, 44, 46, 53, 54, 60, 62, 141
   algebra, 40, 51, 58, 64, 123, 150, 162
   angles, 28, 68, 159
   area, 41, 65, 67, 69, 70, 72, 73, 151
   average, 120, 122, 123, 124, 125, 127
   circles, 27, 34, 71, 73, 74, 76
   data, 113, 116, 118, 121, 122, 123, 124, 126, 127
   diagonals, 141, 142
   diameter, 27, 34, 73, 74, 93
   estimation, 115, 126
   exponents, 55, 59, 140
   fractions, 119, 121, 139, 147, 151, 159
   graphing, 116, 118, 121, 122, 123, 124, 126, 127
   math facts, 57
   mixed operations, 40, 42, 51, 54, 55, 56, 58, 61, 63, 154, 160
   multiplication, 41, 52, 54, 55
   number line, 55, 146
   odd numbers, 61
   percents, 133
   perimeter, 67, 69, 72
   permutations, 124, 155
   place value, 44, 46, 54, 59
   polygons, 34, 68, 69, 70, 141, 142, 144
   probability, 73, 125
   quadrilaterals, 65, 67, 69, 72
   ratios, 134, 153
   rotational symmetry, 82, 83, 85
   skip counting, 48, 52, 158
   slope, 49, 154
   square numbers, 61, 65, 66
   squares, 46, 65, 66, 148
   subtraction, 57
   symmetry, 80, 83, 85, 139
   tangent, 35, 71
   triangles, 28, 69, 71
   triangular numbers, 53
   vectors, 161
   vocabulary, 33
math facts, 57

math history, 110, 114, 115, 116
Math Is Fun, 112, 175
math makers, 4, 20, 93, 149, 168
Math Munch, 112, 175
math phobia, 2, 7
math quilt, 151
math rebellion, 1, 32, 170
math riddles, 159
Math Teachers' Circle, 40
Math Through the Ages, 110, 177, 180
math translation, 103
mathematical reality, 8
mathematical thinking, 166
   Archimedes and, 64
   in the jungle, 8
   noticing, 20
   prompts, 12, 95
   real math, 22, 27
   struggles and, 165, 168
Mathematician Project, 111, 112, 113, 115, 116, 117, 182
Mathematickles, 90, 178
Mathigon, 112, 142, 175
   timeline, 111, 112, 113, 115, 116, 117
maxagons, 144
Mayan numerals, 111
mazes, 82
mean (average), 120, 122, 123, 124, 125, 127
measurement, 118
   benchmarks, 125
   floor plan, 122
   units, 124, 125, 134, 157
Measurement, 8, 180
meme, 162
memory, 172
   building on, 21
   doodling and, 75
   prompt, 97, 106
   school math, 7, 9
   tricks, 108
mental math, 108, 160
metacognition, 12, 26, 94, 95
mindset, 10, 16, 19, 88
mini-zine, 105

misère game, 38
mistakes
   and growth mindset, 88
   good company, 8
   language, 30
   prompt, 97, 99, 103
mixed operations
   brainstorming, 54, 154, 160
   game, 40, 42, 51, 56
   puzzle, 55, 56, 58, 61, 63
mnemonics, 93, 108
money
   Is it math?, 100
   mental math, 108
   puzzle, 102, 131, 153
   spending, 112, 127, 129
   What is it?, 103
Moscovich, Ivan, 146, 181
multiplication, 41, 54, 55
   binary, 113
   math facts, 57
   skip counting, 48, 52, 158
multiplication wheel, 52
Museum of Mathematics, 88
mystery numbers, 150
mystery of numbers, 51

N
National Museum of Mathematics, 88
nature walk, 8, 21, 167
Nim variations, 39, 41, 42, 45, 46, 47, 160
notice and wonder, 20
Nrich Maths, 83, 112, 175, 181
number boxes, 54, 59
number facts, 57
number line, 55, 146
number partitions, 141
number pyramids, 62
number riddles, 159
number snakes, 58
number yoga, 61
Numberless Math Problems, 104, 185
numberstorming, 54

O
obstacle course, 122

octagons, 34, 78
odd numbers, 61
old MacDonald's farm, 157
Ollerenshaw, Kathleen, 64, 181
One Hundred Up, 42
One Right Answer, 22, 27, 171
online resources, 174
open number line, 55
Orlin, Ben, 24, 49, 181

P
painting blocks, 70, 71, 73
palindromes, 146
Parker, Ruth, 16, 182
partitions, 141, 143
patience, 17, 21, 138
payday puzzle, 108
pentagons, 34, 71
pentagram, 71
percents puzzle, 133
perimeter, 67, 69, 72
Perkins, Annie, 164, 182
permutations, 124, 155
phobia, 2, 7
pi, 34
pi poem, 93
Pig, 41
pixel graphics, 77
place value, 44, 46, 54, 59
Place Value Nim, 46
planning, 16, 31
play, 51, 75
   importance of, 5, 9, 35
playing cards, 38, 51, 54
poetry
   acrostic, 94
   haiku, 89
   limerick, 151
   pietry, 93
   square, 87
   the Fib, 92
   tips, 86
   word equations, 90
Polo, Marco, 166
Pólya, George, 137, 182

polygons
    area, 70
    diagonals, 141, 142
    in a hexagon, 68
    math art, 81
    maxagons, 144
    perimeter, 69
    vocabulary, 34
Post, Sonya, 42, 182
Presidents' life expectancy, 113
Pride and Prejudice, 167
probability, 73, 125
problems without numbers, 104
problem-solving
    looking back, 22
    prompt, 95, 104
    seeing and, 20
    struggles and, 109, 165, 168
    tips, 128
prompts, types of, 10, 31, 169
proofreading, 30
proofs, 12, 56, 102
Public Math, 105
punctuation, 30
Pythagorean theorem, 148

Q

quadrature, 73
quadrilaterals
    area, 65, 67
    connecting dots, 69, 72
    puzzle, 161
    vocabulary, 34
quarter the cross, 147
questions
    and proofs, 102
    as a sign of learning, 37
    brainstorming, 20, 25, 96
    evergreen, 169
    geometry, 64
    journalist's, 20, 96
    posing, 3, 93, 137, 164, 169
    problem-solving, 128
    school math, 14
    the answer is, 151

questions (*continued*)
    types of prompt, 11, 13, 169
    value of struggle, 22
quilt fractions, 151
quotation prompts, 13

R

ratios, 134, 153
reason-poems, 102
rebels, 1, 32, 170
recipes, prompts as, 16
rectangles, 65, 67
Reinhart, Steve, 7, 182
Reisch, Gregor, 6, 182
reluctant writers, 18, 25
research
    play as, 5
    questions, 14, 137, 164
    tips, 110
resources, 174
riddles, 159
right answers
    and justice, 171
    answer-getting, 128
    as performance, 14
    how to check, 22, 107
    literal, 60
    shortcut, 168
    side-effect of reasoning, 27
rock-paper-scissors, 134
Roman mosaic, 83
Roman numerals, 111
rotational symmetry, 82, 83, 85
Row Call, 49
rules of math, 9, 19, 33, 51

S

Sallay, Iva, 57, 182
Same But Different, 131, 175
Sanders, Savannah, 51, 183
school math
    discouragement, 2, 7, 170
    follow directions, 9, 19, 51
    performance, 14
    prompt, 100
    speed, 7

science fiction, 103, 130
Scientific American, 112, 178
Secret Number Codes, 40
see, wonder, create, 19, 167, 169
seeing
    and problem-solving, 20, 166
    how to teach, 16, 20
    in geometry, 64
    prompt, 87, 92, 126
self-assessment, 12, 22
senryū, 89
Sequencium, 49
Shah, Manan, 60, 183
sidewalk math, 80
Sierpinski triangle, 70
silly equations, 157
Sim, 43
six-word stories, 87
skip counting, 48, 52, 158
Skip-Counting Game, 48
slope, 49, 154
Slow-Reveal Graphs, 116, 118, 176
Smith, David Eugene, 96, 183
snakes, 58
social media meme, 162
Solve Me, 129, 160, 176
spelling, 30
spirolaterals, 147
squarable numbers, 145
square numbers, 61, 65, 66
square poem, 87
squares, 46, 65, 66, 148
squaring the circle, 73
Srinivasan, Bhama, 75, 183
stacks of paper, 140
stages of growth, 24
Steward, Don, 58, 72, 113, 139, 144, 174, 183
story problem challenge, 149
storytelling
    be an author, 162
    bus, 152
    business tycoon, 159
    favorite characters, 153
    gadgets, 155

storytelling (*continued*)
    movie props, 161
    Old MacDonald, 157
    pets, 156
    restaurant, 150
struggles
    good company, 8
    homework, 7, 167
    value of, 22, 24, 109, 165, 168
    with math, 10, 19
    with writing, 17, 25
Student Math Makers Gallery, 149, 169
Su, Francis, 166, 184
Substitution Game, 42
subtraction, 57
superheroes, 129
supplies, 18
surprise puzzle, 107
swap equation, 139
symmetry, 80, 83, 85, 139
symmetry puzzles, 80

T

tablecloths, 83
tangent, 35, 71
tangrams, 142
Tanton, James, 1, 184
task card books, 31
taxicab geometry, 138
tessellations, 78, 83, 148
testing, 7, 103, 106
thesis statement, 110
Thompson, Christy Hermann, 118, 184
threeven numbers, 163
tic-tac-toe variations, 42, 45, 46, 49, 162
times table, 57, 113
tools of learning, 19, 167
translating math, 103
triangle sums, 61
triangles
    area, 158
    equilateral, 71
    Euler puzzle, 142
    in a circle, 28
    prompt, 69, 80, 161

triangles (*continued*)
   right, 28, 72
   Sierpinski, 70
triangular numbers, 53
Tsyanshidzi, 45
twisted cliché, 60
two truths and a lie, 151
types of prompts, 10, 31, 169

U-V
uncurriculum, 17
units of measurement, 124, 125, 134, 157
Vandervelde, Sam, 40, 184
vector algebra, 161
Visual Patterns, 133, 156, 176
vocabulary
   Frayer model, 89
   geometry, 33
   grid vs. array, 33
   number properties, 54
   silly definitions, 87
   the answer is, 151
   threeven, 163
   word wall, 94

W
weather, 124
Wedekind, Kassia Omohundro, 118, 184
What number am I?, 159
What's Going On in This Graph?, 116, 118, 176
Which One Doesn't Belong?, 133, 163, 176
White, Geoff, 172, 185
Wiles, Andrew, 23
Williams, Farrar, 104, 185
wondering, 20, 98, 99
   and questions, 37, 93, 137, 164, 169
word equations, 90
Would You Rather?, 106, 108, 135, 153, 176
Writer's Jungle, The, 24, 177
writing tips, 25, 86, 96
Wythoff's Nim, 47

X-Y-Z
year game, 56
yoga, 61
Zinsser, William, 86, 185

About the Author

FOR MORE THAN THREE DECADES, Denise Gaskins has helped countless families conquer their fear of math through play. The author of *Let's Play Math: How Families Can Learn Math Together and Enjoy It,* Denise has taught or tutored mathematics at every level from preschool to undergraduate physics. She shares math inspirations, tips, activities, and games on her blog at DeniseGaskins.com.

Get More Playful Math

Join Denise's free newsletter list to receive an eight-week "Playful Math for Families" email course with math tips, activities, games, and book excerpts. Also, about once a month, she'll send out additional ideas for playing math with your kids. And you'll be one of the first to hear about new math books, revisions, and sales or other promotions.
tabletopacademy.net/mathnews

Math Games, Tips, and Activity Ideas for Families.

TabletopAcademy.net/MathNews

Books by Denise Gaskins

tabletopacademy.net/playful-math-books

"Denise has gathered up a treasure trove of living math resources for busy parents. If you've ever struggled to see how to make math come alive beyond your math curriculum (or if you've ever considered teaching math without a curriculum), you'll want to check out this book."
— KATE SNOW, AUTHOR OF MULTIPLICATION FACTS THAT STICK

Let's Play Math:
How Families Can Learn Math Together — and Enjoy It

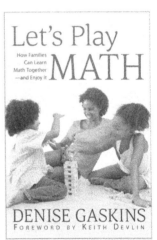

Transform your child's experience of math!

Even if you struggled with mathematics in school, you can help your children enjoy learning and prepare them for academic success.

Author Denise Gaskins makes it easy with this mixture of math games, low-prep project ideas, and inspiring coffee-chat advice from a veteran homeschooling mother of five. Drawing on more than thirty years of teaching experience, Gaskins provides helpful tips for parents with kids from preschool to high school, whether your children learn at home or attend a traditional classroom.

Don't let your children suffer from the epidemic of math anxiety. Pick up a copy of *Let's Play Math*, and start enjoying math today.

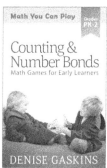 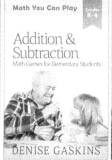 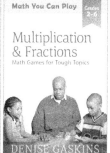 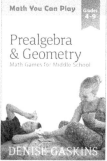

The *Math You Can Play* Series

You'll love these math games because they give your child a strong foundation for mathematical success.

By playing these games, you strengthen your child's intuitive understanding of numbers and build problem-solving strategies. Mastering a math game can be hard work. But kids do it willingly because it's fun.

Math games prevent math anxiety. Games pump up your child's mental muscles, reduce the fear of failure, and generate a positive attitude toward mathematics.

So what are you waiting for? Clear off a table, grab a deck of cards, and let's play some math.

The *Playful Math Singles* Series

The Playful Math Singles from Tabletop Academy Press are short, topical books featuring clear explanations and ready-to-play activities.

Word Problems from Literature features story problems for elementary and middle school students based on family-favorite books such as *Mr. Popper's Penguins* and *The Hobbit*. Step by step solutions demonstrate how bar model diagrams can help children reason their way to the answer.

70+ Things To Do with a Hundred Chart shows you how to take your child on a mathematical adventure through playful, practical activities. Who knew math could be so much fun?

More titles coming soon. Watch for them at your favorite online bookstore.

CPSIA information can be obtained
at www.ICGtesting.com
Printed in the USA
FSHW012303151221
86961FS